ARMCHAIR GEOLOGY

John Priestley

MA (Oxford)
MSC (Liverpool)

Copyright Page

Published by Lulu Books

ISBN 978-1-326-20493-8

© Copyright 2015

First Edition

By the same author:

Tyke on a Bike – the canals of northern Britain, as viewed by a Yorkshireman
History of the British Isles to 1714 AD
History of the British Isles 1714-2010
The Sword and the Claymore – complete political history of the British Isles from the earliest times to 2010, also published as *History of Britain* and *History of England*
Oracle e-Business Consultancy Handbook – Essays, Hints and SQL Scripts for System Users in Manufacturing, Supply Chain and Finance
History of Science
Geography of China
Jeeves and Wooster Short Stories

Contents

ARMCHAIR GEOLOGY 1
Copyright Page 3
Contents 5
THE GEOLOGICAL PERIODS 7
Chapter 1 – The Development of Geology 9
Chapter 2 – Minerals and Rocks 20
Chapter 3 – Modern Geology 38
Chapter 4 – Atmosphere and Ocean 52
Chapter 5 – Vulcanism and Faulting 64
Chapter 6 – The Precambrian 80
Chapter 7 – Lower Palaeozoic 97
Chapter 8 – The Geology of Scotland and the Lake District 113
Chapter 9 – Devonian 123
Chapter 10 – Lower Carboniferous 134
Chapter 11 – Upper Carboniferous 146
Chapter 12 – Permo-Trias 157
Chapter 13 – Jurassic 172
Chapter 14 – Cretaceous 187
Chapter 15 – Paleogene 197
Chapter 16 – Neogene 213
Chapter 17 – Quaternary 225
Chapter 18 – Erosion and Landforms: Humid, Periglacial and Glacial 245
Chapter 19 – Limestone, Coastal and Desert Landforms 258
Chapter 20 – Rivers 272
Appendix I – Mountain Building in the USA 290
Appendix II – The Geological Structure of China 301
Appendix III – Geological History of Australia 314
Appendix IV – Zambia Notes 330
Appendix V – Outline Geomorphology of the Soviet Union 335
BIBLIOGRAPHY 339

THE GEOLOGICAL PERIODS
CENOZOIC ERA
QUATERNARY
.01 MYA (million years ago) – Holocene: Recent
2.58 – Pleistocene: Ice ages

NEOGENE (Upper Tertiary)
5.3 – Pliocene: Red Crags of Suffolk
23 – Miocene: Alps & Himalayas

PALEOGENE (Lower Tertiary)
36.5 – Oligocene: Rapid cooling
---------Eocene mass extinction--------
53 – Eocene: Hot, verdant earth
65.5 – Paleocene: Recovery from catastrophe

MESOZOIC ERA
----------K-T Mass Extinction------------
145 – Cretaceous: Chalk, dinosaurs
199 – Jurassic: Dinosaurs
----------Triassic mass extinction----------
250 – Triassic: New Red Sandstone, rock salt

PALAEOZOIC ERA
----------Permian mass extinction----------
299 – Permian: Desert Britain
----------Permo-Carboniferous Glaciation
----------Hercynian or Variscan Orogeny
359 – Carboniferous: Coal Swamps, Great Limestone
----------Devonian extinction---------------
416 – Devonian: Old Red Sandstone
----------Caledonian orogeny---------------
443 – Silurian
----------Ordovician glaciation and mass extinction-------
488 – Ordovician: Greywackes of Southern Uplands
542 – Cambrian: North Welsh slates, intermittent glaciation

PRECAMBRIAN ERA
2500 – Proterozoic: Torridonian Sandstone
4550 – Archaean: Lewisian Gneiss

Chapter 1 – The Development of Geology

What is geology? It is the study of rocks, everyone knows that, and the first thing it does is to classify rocks into different types. However it does break down into a number of sub-disciplines. The first of these is stratigraphy or historical geology, the study of the geological column of rocks, divided by time periods, extending from the earliest Precambrian formations to recent estuarine deposits still being laid down. Then there is palaeontology, the study of the fossils within the rocks. Many palaeontologists are really zoologists. Next is physical geology or geomorphology, the study of landforms such as hills, valleys and river systems, a subject shared between geology and geography. Physical geology also includes the large-scale processes which have modelled the earth in a fundamental way – plate tectonics, vulcanism and glaciation. Another branch of the subject is mineralogy, the study of the minerals which are found within rocks and their often arcane crystallography, and beyond the scope of this book. A further subdivision is geological mapping, both of the surface and subsurface rocks, of obvious importance in mineral exploitation, especially when drilling for oil! Again, this book will not attempt to explain the interpretation of geological maps, but anyone looking at these rather specialist documents will be assisted by a knowledge of the stratigraphy which is contained in these pages. There are also relatively new disciplines within geology such as geochemistry and paleomagnetism.

In terms of the sciences, geology is middle-aged. It is nowhere near as old as the first real science, astronomy, but much older than say genetics, which only really dates from the start of the twentieth century. The first person to think scientifically about the subject was the Dane Nicholas Steno, who published a book in 1678 proposing fundamental rules about how (sedimentary) strata form. He said that rocks accumulated one on top of the other over long distances, and if they are not horizontal, they must have been tilted after deposition. If they end abruptly in a cliff, they must have been eroded away. Also if sea shells are found on top of a mountain, then the sea level must once have been much higher, or the land surface lower.

The first important geological publication was *The Theory of the Earth* by James Hutton (1726-1797), which appeared in 1788 and in its full version in 1794. Until the time of Hutton, it was widely accepted that the earth was created in 4004 BC. Hutton realised that this could not possibly be so. He saw that the earth's crust was recycling itself, and that the constant process of erosion of the land into the sea meant that land must also rise by uplift, that indeed the sea floor could form mountains in the next cycle. In his famous words, he saw "no vestige of a beginning, no prospect of an end." He was the first pioneer of the concept of deep – geological – time; but he had no idea how deep. He also saw no need to invoke catastrophic processes in the creation of the landscape as he saw it. If enough time were allowed, processes which could be seen operating today, including normal erosion and vulcanism, could explain everything: *the present is the key to the past.*

In the next generation after Hutton great strides were made in the establishment of the sequences in the English stratigraphical column and the fossils it contains. The most important name from this era is William Smith (1769-1839). A man of humble origins – his father had been a blacksmith – Smith worked as a surveyor of canals and coal mines in the Somerset area. In 1799 he published the world's first real geological map, of the area around the city of Bath. By dint of much hard labour and often with little money – he overstretched himself and ended up in a debtor's prison – he eventually published the first geological map of England, Wales and the southern part of Scotland in 1815. We will learn more of William Smith in due course. Also active in this period was Mary Anning, the famous collector of Jurassic marine fossils – fantastic specimens of plesiosaurs and ichthyosaurs. She worked in the cliffs of the Lyme Regis area on the south coast of England, loose shales (Blue Lias) which were continuously pounded by the sea to reveal fresh specimens. She is remembered by the phrase "She sells sea shells by the sea shore"! However, she had stumbled upon what every fossil hunter needs – an extensive exposure of sedimentary rock, bare at the surface, and hard to find in temperate areas. Grassy fields with a thick layer of soil are no use for fossil hunting. In the course of this book many famous fossil sites will come to light which have been in deserts such as Fayum in the Sahara, Hsanda Gol in the Gobi desert, and the western deserts of the United States; in the bare Arctic (Greenland, Ellesmere Island); and in quarries and along coasts.

Field workers enjoying the **Blue Lagoon**, kept warm by hot rocks near the surface. Iceland is the world's most geologically-influenced country

The books of James Hutton proved so impenetrable that the real founding publication of geology came to be seen as the work of another man, Charles Lyell (1797-1875), nearly forty years later. He codified much of modern geology as it is understood today when his *Principles of Geology* appeared in three volumes in the years 1830-3. This was one of the most influential books taken by Charles Darwin on his voyage on the Beagle, which began two years later. Lyell developed Hutton's idea that "the present is the key to the past", the idea being that geological remains from the distant past can, and should, be explained by reference to geological processes now in operation and thus directly observable. Lyell interpreted geological features as the result of the steady accumulation of minute changes over enormously long spans of time. This is called Uniformitarianism, the assumption that the same natural laws and processes that operate now have always operated in the past, and function at similar rates. The term *uniformitarianism* itself was coined by William Whewell, who also coined the term *catastrophism* for the idea that the Earth was shaped by a series of sudden, short-lived, violent events. In Lyell's time, there was thought to be one major catastrophic event – the flood of Noah, as described in the Bible. In the face of the evidence, some scholars were prepared to accept more than one flood! This idea is also called Neptunism. In modern times it is sometimes difficult to conceive of the opposition people like Lyell and Darwin faced from fundamentalist Christians, who were in those days not freakish Texans but who actually represented mainstream thinking. In any event, uniformitarianism has remained a key principle of geology, but modern geologists, while accepting that geological processes have operated across deep time, no longer hold to a strict gradualism. There can be catastrophic events as well – most famously, large meteors crashing into the earth. Charles Darwin, who was a geologist before he was a biologist, was to apply a similarly uniformitarian view to evolution.

Geologists today apply the principle of uniformitarianism rather as other scientists apply the principle of parsimony, also mysteriously known as Occam's Razor. This states that a simple explanation is always to be preferred to a complicated one – it is the most "parsimonious". When a new theory comes along in geology, the more it invokes processes which cannot be seen operating on the earth today, the more scepticism it provokes. We shall see how this works later in the modern idea of the Snowball Earth, which proposes sea ice at the equator. Very much the same thing happens in physics when new ideas come forward which require the suspension of the Second Law of Thermodynamics, which states that systems become more disorganized without an external input of energy – the universe is wearing out. If your new theory does not fit within this law, think of another one!

Principles of Geology was first published shortly before another major breakthrough in the field. In 1837 a Swiss academic called Louis Agassiz (1807-1873) became the first scientist to propose that the Earth had been subject to a past ice age. His attention had been drawn to the presence of erratic boulders, rocks far away from their original outcrop, by one Jean de Charpentier. Several walkers had by then arrived at the conclusion that the erratic blocks of alpine rocks scattered over the slopes and summits of the Jura Mountains – which contain no glaciers – must have been moved there by ice. In 1840 Agassiz published a work in two volumes entitled *Études sur les glaciers* ("Study on Glaciers"). In it he discussed the movements of the glaciers, their deposits (moraines and erratics), their influence in grooving and rounding the rocks over which they travelled, and in producing the striations and *roches moutonnées* (boulders which are smooth on the upstream side, but plucked out on the lee side) seen in Alpine-style landscapes. He not only accepted Charpentier's idea that some of the alpine glaciers had extended across the wide plains and valleys drained by the Aar and the Rhône, but he went still farther. He concluded that, in the relatively recent past, Switzerland had been another Greenland. Instead of a few glaciers, one vast sheet of ice, originating in the higher Alps, had extended over the entire valley of northwestern Switzerland until it reached the southern slopes of the Jura, which, though they checked and deflected its further extension, did not prevent the ice from reaching the summits of the range.

Agassiz travelled to Scotland with the English geologist William Buckland, in 1840. The two found in different locations clear evidence of ancient glacial action. The mountainous districts of England, Wales and Ireland were also considered to constitute centres for the dispersion of glacial debris; and Agassiz remarked "that great sheets of ice, resembling those now existing in Greenland, once covered all the countries in which unstratified gravel (boulder clay shunted by ice) is found."

Glacier descending **Mont Blanc** in France in the heart of Agassiz territory

Two more figures from the nineteenth century of importance to geology might count as zoologists or anatomists rather than geologists. One of these is Charles Darwin himself (1809-82), whose *Origin of the Species* was published in 1859. A contemporary was Sir Richard Owen (1804-92), the driving force behind the establishment of the Natural History Museum in South Kensington, still the top choice for any child or adult on a visit to London and famously home to *Diplodocus* and many other dinosaurs. It was Owen who coined the very term dinosaur ("terrible lizard"). He also made a remarkable discovery about the anatomy of tetrapods – creatures with four limbs; that all their limbs follow the SAME blueprint. In the case of the human arm, this is one large upper bone (the humerus) which then articulates to two bones (ulna and radius), which then attach to a series of small bones at the wrist, before spreading out into five digits or fingers, each made up of several separate bones. The human leg follows the same pattern. Owen, however, noted that this system of bones was exceptionally similar in all tetrapods, be they frogs, birds, pterosaurs, bats, lizards, seals, theropod dinosaurs (upright carnivores) or even humpback whales. He published his findings in a book called *On the Nature of Limbs* in 1849, ten years before *Origin of the Species* appeared. Darwin, far from being ahead of his time, was only just in time with his theory. The common architecture of limbs points to one, now very obvious conclusion – that all tetrapods evolved from a single common ancestor. What that ancestor might have been, we shall consider when we look at the Devonian period.

Origin of the Species did not create the idea of evolution, which was already well established, but a mechanism for it, natural selection, or more specifically the competition between different members *of the same species* for survival – "survival of the fittest" (not Darwin's own words). Darwin's idea was that minute (genetic) differences between the offspring of any species, as brought about by mutations, could confer an advantage in the highly competitive environments in which most creatures live. In that case the mutation would survive and prosper in future generations. If such a mutation had a deleterious effect then its bearers would not survive to reproduce successfully. As tiny increments of mutations would take many generations to become established, it was clearly going to take a vast amount of time to produce all the creatures on the Earth, let alone all the extinct creatures identifiable from the fossil record. Yet, fundamentally, it was such a simple idea.

Natural selection not only affects the ability of an animal to feed itself and survive predation. It needs to reproduce as well, and here Darwin noted some of the truly remarkable features produced by sexual selection which would otherwise seem a distinct disadvantage, most famously the tail of the peacock. Nobody really knows why the dowdy peahens like mates with such splendid feathers, but clearly, they do, even if it means that their mates can barely fly, and thus find it much more difficult to escape from hungry leopards. Again, the males of most deer species have quite ridiculous headgear, with which they batter each other – sometimes to death – in the mating season; but without it, they do not reproduce.

Darwin had been influenced by Thomas Malthus who in 1798 published *An Essay on the Principle of Population*. In this he stated that human populations increase exponentially, and could double with every generation, whereas food production can only increase at lower, arithmetic rates. Therefore sooner or later the population would have to be reduced by famine, warfare or disease. Darwin saw that the same pressures of food supplies and predation applied to animal species. If a stable numbers of any animal species were to be maintained, many would have to die without reproducing first, as birth rates generally greatly exceed the rates needed merely to maintain the same population.

In fact Darwin had already observed the amazing changes brought about by the selective breeding of pigeons and other domesticated species, and he saw that natural selection operated in the same way. In other words Darwin was looking at what we would call genetic changes brought about by selective breeding, either natural or artificial. Small changes occurring naturally between one generation and its offspring could in the end produce whole new species, as had clearly happened with the finches of Galapagos.

In fact Darwin had become an expert field geologist whilst in South America (taking time off from the voyage of the *Beagle*). Indeed he had spent three weeks with the famous Cambridge geologist, Adam Sedgwick (of whom more later) in North Wales before embarking on his travels. As Darwin's new theory more or less dispensed with the need for a Creator, it set off the greatest bombshell in all Victorian science, affecting geologists as much as biologists. Sedgwick was disgusted by the new idea. After publication Darwin had to face the music, tormented as he was by his own theory – "like confessing to a murder". The first edition of his new book, 1,250 copies, sold out on the first day, and soon the knives were out. However Darwin did to some extent leave it to others, notably Thomas Huxley, to state his case in public. In a famous debate held in the Oxford Zoological Museum on 30 June, 1860, Huxley took on Samuel Wilberforce, the Bishop of Oxford. After Wilberforce demanded of Huxley whether his ancestry from the apes came by way of his grandmother or grandfather, the meeting collapsed in tumult, both sides claiming victory.

If evolution happened by a series of small steps, the opponents argued, then where are the intermediate species? In fact the geological record IS short of intermediate species, simply because the preservation of fossils is such a haphazard process. Bats, for example, first appear fully formed, with no hint as to how they got to fly. The most famous intermediate species, *Archaeopteryx*, part bird and part reptile, was actually discovered in Bavaria in 1861, and might have helped Huxley in the debate a year sooner. Even today, this debate still goes on. Some palaeontologists see evolution not as a steady process, but coming in sudden bursts – punctuated by periods when nothing much seems to change.

Also, even Darwin conceded that his process of natural selection required enormous amounts of time – deep time – to work. As late as 1900, physicists such as Lord Kelvin were prepared to give the Earth an entire history of no more than 100 million years, yet it was known that the large thicknesses of older strata (the Precambrian) contained no sign of life, so the timescale simply looked too short. Darwin found himself very much on the defensive, despite the rather obvious similarities between people and chimpanzees!

****

A good place to see geology in action is Finland. The country is part of the Baltic Shield, Precambrian rocks of immense age. These can be seen in the road cuttings, many of them red metamorphic and igneous rocks. There are no mountains, the hummocky landscape rolling on to Lapland in the far north. The land has thousands and thousands of lakes. Rain and snowfall is not heavy, but there is little evaporation so this is a very wet place. In fact, there is so much water that I have the feeling that should the climate ameliorate (as predicted by global warming), then the Finns will have difficulty in keeping out Johnny Foreigner. Up to now, they have kept him out very successfully, and there seemed to me to be a smaller proportion of foreigners here than anywhere in northern or western Europe. But after all, who would want to live in Finland?

I spent over 18 months in Finland itself from 2005, though I flew home most weekends. It is a fascinating country for someone like me who had never spent any time in the far north. The work was based at a town of about 50,000 inhabitants called Hameenlinna, an hour's drive north of Helsinki. I arrived very late each Monday evening, often after midnight local time, though local time was two hours ahead of UK time. The drive northwards was instructive. The forest starts at Helsinki airport itself. Though there are cleared areas, the forest rolls on for hundreds of miles northwards, the cleared areas diminishing all the way. You have the feeling that the farmland has been cleared for now, but the forest could take it back any time.

Geographically and geologically, Finland is like a small version of Canada. As well as the Precambrian shield rocks and the boreal coniferous forests, both countries are dominated by thousands upon thousands of lakes. Hameenlinna was built across one – it froze to a depth of about one metre (three feet) every winter. The ice from the last ice age must have retreated about 10,000 years ago, but it left a mess behind it, and nature can only work slowly to repair the damage in places like this. Soil formation, for example, happens much more slowly than it does in Britain, and many bare rock surfaces remain. Also it is clear that the ice sheets completely disrupted the drainage system. The country has lots and lots of surface water, but can you name me one Finnish river? Even now I cannot name one; each is just a short stretch of water before the next lake, or the sea.

In addition, the whole country and the seas around it were depressed by the weight of the Pleistocene glaciers and rose as much as 180 m (600 ft) due to post-glacial rebound over the late Pleistocene and Holocene, and are still rising today. An extreme uplift of 270 metres (900 feet) has been measured in the northern part of the Gulf of Bothnia (Finland), which continues at the rate of about a centimetre a year.

After my first winter there, I couldn't imagine why the Finns did not emigrate en masse to some warmer place. The snow settled permanently in November and stayed till the middle of April, having become dirty and unattractive months earlier. As a matter of fact, the winter is dominated by high pressure, and much of the snow falls at the beginning and end of it, and little in the middle. At some point every year, in January or February, the temperature falls to 25 degrees below zero (Centigrade), or lower. In this boreal chill I felt as if I might as well not be wearing clothes if they were only one layer thick (like trousers).

But of course, this is Finland, and the Finns are used to it. They dig their plumbing very deep and they even have municipal central heating. All the houses are well insulated and have triple–glazing in the windows. Whilst there I became friendly with my regular taxi driver, Mikka.

"How do you manage for fuel when it gets this cold?" I asked.

"Petrol doesn't freeze, even at minus 30," he replied. "And we have three types of diesel – summer diesel, winter diesel and then Arctic diesel!"

There are other adaptations to the cold. Every car owner has two sets of tyres, summer and winter, and there is a coordinated switchover when the cold sets in. Cars can plug into a heating system in car parks to keep them warm when left out all day. Still the Finns like the sun as much as anyone, and often take tropical holidays in January or February, preferring to disappear into the vast Finnish forests with a fishing rod in summer.

As well as being freezing cold in winter (and not very warm in summer), there is also the daylight issue. At Hameenlinna in mid–winter, it would not be truly light until 10 am, and it would be dark by 3 pm. In summer it was of course the exact opposite and you could open the curtains at 3.30 am to see blazing sunlight.

But funnily enough, the Finns like the cold, or at least, some of them do. In my second year the snow failed to arrive on time. Jukka the sales manager complained bitterly:

"This is no good, neither summer nor winter. You can't do anything! It's so dark!"

The snow cover lightens the gloom of midwinter, and when it falls the locals can go snowboarding and skiing. However, the lady drivers did not share this attitude.

It is well–known that Finnish is the most impenetrable language in Europe. After a couple of months, I had only learnt ten words of it, and then I found out that two of these were actually Swedish – Osterrut and Westerrut. I used to ask the Finns why they spoke this ridiculous language. After all, the Scots and the Irish had their own languages, but they abandoned them in favour of English centuries ago. No one seemed to know the answer to that. Finland was a Swedish possession until 1815, so why didn't they speak Swedish instead? I think the answer to that one is that they probably did, in 1815. Then the Russians took over, until 1917. Even though it falls on December 6, the Finns take their independence day celebrations very seriously, and as they are celebrating independence from the Russians, who can blame them?

President Chirac once observed that the food in England was nearly as bad as in Finland, and in this, I have to agree with the second part. I can't understand how anyone can eat the sort of food that was swallowed with relish by the locals in the factory canteen. There is one thing about this food, however. It is not fattening. The Finns became worried about obesity years ago, and all institutional food is like this. For example at the Honeywell canteen in Scotland, chips were on the menu every single day. I only ever saw these once in the factory canteen. However, there was something I could eat. Every Thursday, pea soup was served, with bread and butter, then pancakes and jam. I had a colleague called Vesa who had been in the army (national service is compulsory) and he explained to me that this is actually the army menu!

The Finnish people nearly all look like each other's cousins and I found it easy to "spot the Finn" on the flight from Manchester. I think the reason for this is that there was only a small population a thousand years ago, when agriculture began in this benighted country. The present population of five million has descended from this small gene pool, so they are in fact each other's cousins.

As I have found throughout my working life, the places with the worst self-image are also the most welcoming, and the Finns were certainly that. They actually looked pleased to see this Johnny Englishman when I arrived each week.

Chapter 2 – Minerals and Rocks

Before starting to look at the stratigraphic column, it would be as well to take a look at the structure of the earth, the rocks themselves, and the minerals they contain. Seismic data gathered from earthquakes reveals the fact that there are three fundamental divisions within the earth itself: a thin crust on top; beneath it, the mantle; and at the centre, the core. The controlling factor is density, which increases with depth. The rigid top layer composed of the crust and the uppermost part of the mantle is known as the lithosphere. They are separated by the Mohorovičic discontinuity, where the velocity of seismic compression waves suddenly increases, reflecting the greater proportion of iron and magnesium, and the smaller proportion of silicon, found in rocks at that level. This is about 40 km/25 miles down.

Beneath the lithosphere is a viscous layer known as the asthenosphere, 100-200 km (62-124 miles) down, across which the tectonic plates into which the crust is divided can slide. It is also the source of isostatic or relevelling adjustments where material can move aside to accommodate the deep roots of mountain chains. Although we can never see it or go there, the mantle as a whole is very important to us. It has a high level of thermal energy, caused by the decay of radioactive materials and by gravity differentiation within it. This energy gives rise to earthquakes, plate tectonic movements and vulcanism, and it also transforms rocks which dip into it by metamorphism, discussed below. The mantle and the core between them also generate the earth's magnetic field, which shields the surface of the earth from harmful solar rays. These processes are described as endogenous, because they take place within the earth. The processes which arise at the surface of the earth, such as weathering, are described as exogenous.

Layers of atmosphere (top two) and earth:

Stratosphere
Troposphere

==========
Lithosphere
Asthenosphere
Lower Mantle
Core

    All rocks are composed of minerals, and it is the type of its minerals which gives any rock its physical properties. The individual minerals within a rock are most easily seen within coarse-grained igneous rocks, notably amongst the granites which are used to make the decorative facades of public buildings. A typical granite contains three main minerals – quartz (silicon dioxide) (clear), mica (black and shiny, or white) and feldspar (white or pink). Quartz is a very common mineral which also exists in gem form, when it is called chalcedony. This includes various stones such as amethyst, agate, onyx, jasper, carnelian, bloodstone and petrified wood. The most common varieties of mica are muscovite (white) and biotite (black), and it is biotite which features so strongly in granite. Feldspar minerals themselves form into two main groups, orthoclase (or potash feldspar, found in acidic rocks such as granite) and plagioclase, containing calcium and sodium instead of potassium, and found in basic rocks. Granites also often contain porphyries (usually feldspars), much larger mineral crystals – sometimes inches across – which have crystallised earlier than the mass around them. Associated with granites are veins of white quartz. This mineral has a low melting point and is often dissolved around granites, travelling upwards into the country rock as a superheated stream, then cooling and recrystallising.
    Other common mineral found in igneous and metamorphic rocks are black or green hornblende, olivine and pyroxene. Hornblende is one of a group of minerals known as amphiboles. Olivine takes its name from its olive colouring, which comes from iron in its ferrous state. Clear, gem quality olivine is known as peridot. A common type of pyroxene is known as augite.
    Mineral salts can also be found in crystalline form. One such is selenite, which is in fact gypsum (hydrated calcium sulphate). The most enormous crystals of this substance, the largest 11 metres (36 feet) in length and 4 meters (13 feet) in diameter have been found in an underground cave in Mexico, Cueva de los Cristales. They owe their existence to superheated water, charged with minerals, emanating from a magma chamber below the cave. In mineralogy, ancient and modern, the term "spar" is used loosely to describe any non-metallic mineral

with readily discernible faces. It is derived from the same word as spear, and is particularly relevant to a mineral like gypsum, where the crystals do resemble spears. (The compact, non-crystalline form of gypsum is alabaster.) Another "spar" is fluorspar or fluorite, calcium fluoride which comes in a variety of colours, notably blue when it is called Blue John in Derbyshire

Only about 3,760 types of mineral are known to exist on earth (by some classifications, only half that number). One reason for this rather small number is that only eight elements make up ninety percent of the earth's crust – oxygen (46.6% by weight), silicon (27.7%), aluminum, iron, calcium, sodium, potassium and magnesium, in that order. Silica (silicon oxide, SiO_2), in the form of sand or quartz, is of course very common, as are silicates where it is compounded with iron, aluminium and other metals. Most common rock types apart from carbonates (chalk, limestone) are combinations of different silicate minerals, so silicon performs a role in geology analogous to that played by carbon in the organic world – but it is far from being as versatile as carbon, reducing the range of options for creating new forms. In quartz, silica pyramids bond in all directions, creating a mineral without natural breakage planes. Mica has silica pyramids as well, but they are bonded only into sheets, not three-dimensionally, and so easily split apart.

Except in the form of carbonates, carbon does not play so central a part in inorganic chemistry as it does in organic chemistry, where its ability to combine with other elements leads to the creation of huge organic hydrocarbons. This is because it does not combine with metals. Metals themselves prefer to form ionic bonds (cation to anion, normally metal to nonmetal, for example sodium chloride). Carbon prefers to form covalent bonds (sharing electrons) such as methane (CH_4) where the four outer electrons of the carbon atom are shared with four hydrogen electrons.

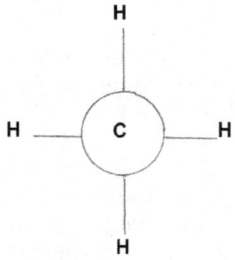

The simplest possible hydrocarbon, methane, where one carbon atom shares its four spare electrons with four hydrogen atoms. Carbon can form whole chains of organic chemicals in this way, a process known as concatenation.

Minerals can of course be valuable in their own right, but only where they have formed under unusual conditions, making them rare. One of these is a form of the common mineral corundum, a simple oxide of aluminum and normally of no special interest. However when it crystallizes deep underground with an admixture of impurities (in fact chromium oxide) to induce colour, it becomes that most expensive of all gems, the ruby. When coloured with titanium it becomes a sapphire. The second-lightest of all the metallic elements, beryllium, can form a very hard mineral called beryl. If transparent this is called aquamarine, and if green, emerald. The most famous gemstone of all, the diamond, is of course nothing but pure carbon, but crystallised at temperatures and pressures which can only be found 150 kilometres (95 miles) down in the mantle of the earth. In fact the mantle temperature found at this depth is thought much too hot for the creation of diamonds, which instead are believed to have formed at the base of cratons, the early continental land masses which sank deep roots into the mantle, as much as three billion years ago.

Another valuable pure element is gold, a metal whose reluctance to combine with any other element means that it can be found in nature, pure and unalloyed. The fact that it does not need to be smelted gave it value to human cultures before such a process was invented – gold has had a special place for thousands of years. It was valued by the Incas whose forefathers crossed the Bering Strait 15,000 years ago. It can be mined, or it can be panned from stream placer deposits. It is sometimes found naturally alloyed with silver, and in fact it does combine with another rare element, tellurium, to form gold telluride. At Kalgoorlie in Australia, the miners used the ore to build roads and houses before, after three years (1896), it was found to contain 14 kilograms (500 ounces) of gold per ton – a figure any modern miner would drool over. The gold could be obtained by simply heating the unpromising-looking ore, which gave off a dense smoke of tellurium to leave a shiny gold-silver alloy behind. Gold also forms natural "salts", compounds also containing sodium and hydroxyl ions. The gold mine at Cripple Creek, Colorado, once the largest in the world, had deposits of this type.

Silver is frequently recovered from other ores, notably the lead ore galena, rather than from its own silver minerals, but these do exist. The most famous silver deposit in the world is at Potosi in Bolivia. Here is a volcanic peak rising to 5,000 m/16,000 feet and riddled with silver ores, mostly cerargyrite. The ores have been mined since 1544.

Other very important minerals are haematite and magnetite, the main ores of iron. Iron itself has two common forms, ferrous and ferric. When combined with oxygen, the formulae are:

Ferrous Iron: $Fe_2 O_3$ – occurs naturally as haematite
Ferric iron: $Fe_3 O_4$ – occurs naturally as magnetite

(Magnetite can contain both ferric and ferrous iron.) The characteristic red coloration of rocks (such as the New Red Sandstone) and soils is often due to the presence of ferric iron. Ferrous iron can give a greenish coloration, notably in the Greensand rocks of the Weald of Kent and Sussex. Other important mineral ores include cinnabar (mercury), galena (lead), cassiterite (tin) and sphalerite (zinc). There are three important ores for copper: malachite (green), azurite (blue) and cuprite (ruby red). Two sulphide ores of arsenic are known as orpiment and realgar. Deposits of these can be found at hot springs, as in the Yellowstone National Park All of these ores were known to, and used by, the ancients. Indeed the British Isles were called the Cassiterides by the Greeks as the mines of Cornwall were the source of tin for the Bronze Age.

By contrast, the importance of the principal ore of uranium, pitchblende, was not realised until after the discovery of radiation by the Curies at the beginning of the twentieth century. Up until then, the silver miners of Joachimsthal, in the Czech Republic, had cast it aside as useless dross. Another ore with a short history of usage is bauxite, the only commercially useful ore for aluminium. The metal can only be recovered by electrolysis and so aluminium refineries are always situated near cheap sources of electricity. There is a large one on the coast of Iceland, situated there to take advantage of electricity generated from thermal springs.

****

One of the best-known rocks in the whole of the British Isles is the Carboniferous Limestone. It forms the foundation of two of the mast attractive national parks, the Peak District of Derbyshire the Yorkshire Dales.

I have had some encounters with the limestone country of the Yorkshire Dales myself. One of these happened when I was only sixteen years old. Along with two friends, Philip and Poffles, I once went camping at Cracoe in Wharfedale, in the Yorkshire Dales National Park. (There is a very large quarry just nearby.) We had walked ten miles or so across country to get there – it was a lovely day and the Dales looked green, lush and beautiful as always. Part of this traverse included crossing the river Wharfe in an area where there was no bridge. We had no choice but to walk across, first removing our footwear, and carrying our heavy packs on our backs. I was the first in. It was Easter, but there was still snow about on the hilltops. The water – really this was snow, recently melted – felt excruciatingly cold. I also learnt an early lesson about the life cycle of a river. The pebbles on the bottom of it were sharp and slippery, not rounded as they might be further downstream, or on a beach. Towards the middle, the stream suddenly deepened and the water came over my knees, and the other bank looked further and further away. It was agony as I slipped and staggered from stone to stone, at times trapping my toes in the gaps between two rocks, and all the while becoming colder and number in the parts exposed to the water. But I made it across in the end, still having to pick my way carefully over the last, murderous few yards, finally crashing onto the far bank in relief.

I then had the pleasure of watching Poffles and Philip attempt the same thing. Half–way across as he hit the deeper, faster water, Poffles' face began to express alarm. As he ploughed on, swaying violently from side to side or lunging forwards as his feet hit the sharp edges or plunged into holes, he began to curse, the expression on his face reflecting at once frustration and agony and the air turned blue.

The worse things became for him, the funnier it got for me, until I lay on the bank, howling with laughter. I can't ever remember finding anything so funny.

"Will you shurrup! It is not funny! Ouch!"

They made it in the end. We pitched camp and were later joined by my sister Cathy and one of her girlfriends, Leslie, who opened a huge, expensive–looking tin of Frey Bentos stewing steak. You can no longer buy this, but it was rather fatty and full of sweet meat, and it tasted divine. Food always does taste like that when you haven't had any for 12 hours.

****

It is widely known that rocks fall into three great groups – igneous, sedimentary and metamorphic. Igneous (from the Latin for "fire") rocks are created by the hot processes within the earth and have formed from the solidification of molten rock or magma. They in turn divide into two main groups. The first of these is the plutonic or intrusive – welling up under the surface of the earth, cooling slowly and only being exposed at the surface by later erosion. The best-known rocks of this type are granite and gabbro. The hot magma from which these rocks form can itself be composed of the molten remains of other types of rock, deriving from the continuous process of recycling to which the materials of the earth are subjected. The slow cooling of intrusive rocks is reflected in their coarse-grained structure where it is possible to distinguish individual minerals with the naked eye. Characteristically, granite is found in plutons, underground masses which over the eons have in many cases now been exposed at the surface. These form the cores of ancient mountains. Often separate plutons merge at depth to form what is called a batholith, a largely submerged mass of granite of unknown depth.

Incidentally, professional geologists very much dislike the term "black granite", often used, for example, to describe expensive kitchen worktops. If it is a coarse-grained, dark igneous rock, then it is likely to be a gabbro. Granite is not dark, but mottled grey, or pinkish (one well known pink type is the Peterhead granite of Scotland). Another type of plutonic rock is syenite, very coarse-grained and containing large feldspar crystals. It is similar to granite but has little or no quartz. A variety of this, known as larvikite, from the town of Larvik in Norway, can be found on bank and shop fronts the length and breadth of England, grey or greenish in overall colour.

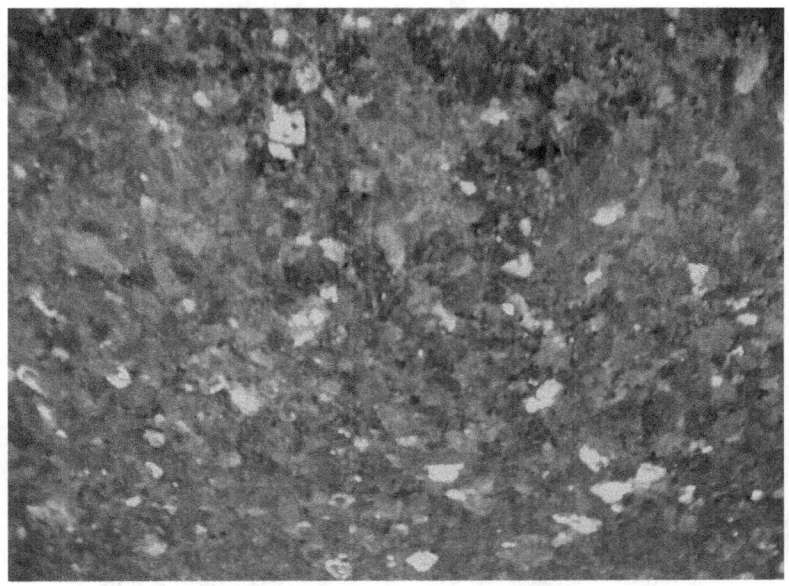

Dark larvikite

Pegmatites are generally types of granite which have cooled very slowly into supersized crystals, 2.5 cm (an inch) or twice that across, sometimes containing much larger individual crystals.

The second great igneous group is the volcanic rocks, extruded directly onto the surface of the earth from volcanoes. These include the solidified magma itself and various other products of volcanoes. Solidified lavas include basalt (a dark, fine-grained rock, well-known from it outcrop at the Giant's Causeway in Antrim and in Iceland), andesite and rhyolite. Basalt is a dark, basic rock, as its name implies, and its presence indicates relatively peaceful eruptions at mid-oceanic ridges. The plagioclase feldspars within it contain calcium (which is not found in the orthoclase feldspars of granite). On weathered outcrops the result can be a lime-rich soil and healthy plants.

Andesite is intermediate in composition (between basalt and rhyolite), and rhyolite is felsic (acidic), which means it contains a lot of feldspar and silica. Both of these erupt much more violently than basalt, especially rhyolite, which is very viscous and jams volcanic vents before exploding. The presence of either of these rocks in the geological record indicates a subduction zone in a closing ocean, that is, a place where heavy oceanic crust is diving beneath the lighter continental crust. Rhyolite is produced by the melting of felsic crust material and it is characteristically pale-coloured, even reddish. It is the fine-grained equivalent of granite. Andesite is produced by the mixture of melted felsic crust and liquid basalt and it is grey in colour. The Andes, thrown up where the Pacific plate is dipping below the South American continent, are andesitic mountains.

Other volcanic material includes pumice, volcanic glass or obsidian and the pyroclasts ("fire fragments") which form a rock called tephra or tuff when blasted out from catastrophic explosions. A very explosive pyroclastic eruption such as that of Vesuvius in 79 AD produces a mass of fiery particles which settle down into a rock called ignimbrite, a word derived from the Latin terms for "fire" and "rain". Like sedimentary rocks, and unlike other igneous rocks, tuff does not have a crystalline structure. Obsidian cools so quickly that no crystallization has time to take place. The result is a perfect natural glass which was much valued in the stone ages for axe heads and cutting tools.

Another igneous formation is the dyke, a relatively narrow band of rock shot vertically through the existing country rock, which is always completely different in geology. These dykes can be very deep. The study of a small-scale geological map may show whole swarms of dykes coursing away from volcanic or granitic centres for dozens of miles. One such swarm heads south-eastwards from the southern part of the Hebrides, in Scotland, which is a volcanic province from the Tertiary. This type of injected rock can also be found in the horizontal rather than the vertical plane, when it is known as a sill. In this case it finds its way between strata of older rocks, often sedimentary, which offer weak bedding planes. Dykes and sills are often composed of dolerite, a rock with the same chemical composition as basalt, the crystalline structure of which is too fine to see with the naked eye. It is a dark rock which weathers out to orange in surface exposures due to the presence of iron. Its feldspars are the plagioclase type and may be visible as pinkish or grey crystals. Another name for dolerite is diabase.

Igneous rocks can be classified on chemical lines into acidic, intermediate, basic and ultrabasic. This classification has nothing to do with the Ph values used in chemistry, but are instead based on the amount of silica in the rock – acid, around 75%, intermediate, about 55%, and basic, 45%.

Acidic rocks (granite, rhyolite) contain a lot of silica and orthoclase feldspar. On modern geological maps, the term "acidic" has now been replaced by "felsic" (feldspar plus silica). Intermediate rocks (andesite and its coarse-grained equivalent, diorite) include plagioclase, hornblende and/or pyroxene. Basic rocks (basalt, gabbro, dolerite) have much less silica, plus plagioclase feldspars and a lot of magnesium and iron. Again, the term "mafic" (magnesium plus ferric) has replaced "basic" on modern maps. Ultrabasic rocks have a different mineral mix. One of the most notable is peridotite, composed mainly of the heavy minerals olivine and pyroxene, and thought to have originated in the mantle – the region beneath the crust, 30 kilometres (18 miles) or more below the surface of the earth. Despite its exotic origin, it is not uncommon in shop frontages. It is characteristically green or brown and is always riven by lighter-coloured veins. An associated rock is kimberlite, which forms pipes diving deep into the earth in southern Africa, the mother lodes for diamonds. Another rock from the great depths is eclogite, which contains crystals of garnet and which is also put to decorative use.

In practice, ground exposures of coarse-grained acidic rocks (the various types of granite) are much more common than outcrops of the intermediate and acidic types, diorite and gabbro.

One way to detect the presence of mafic rocks in the subsurface geology is by their effect on gravity. As they contain relatively large amounts of heavy manganese and iron, any item will actually weigh slightly more in the presence of significant rock formations of this type. It is not the sort of difference that can be picked up on a bathroom scale, but it can easily be detected on a gravimeter, an instrument designed to pick up precise differences in gravity.

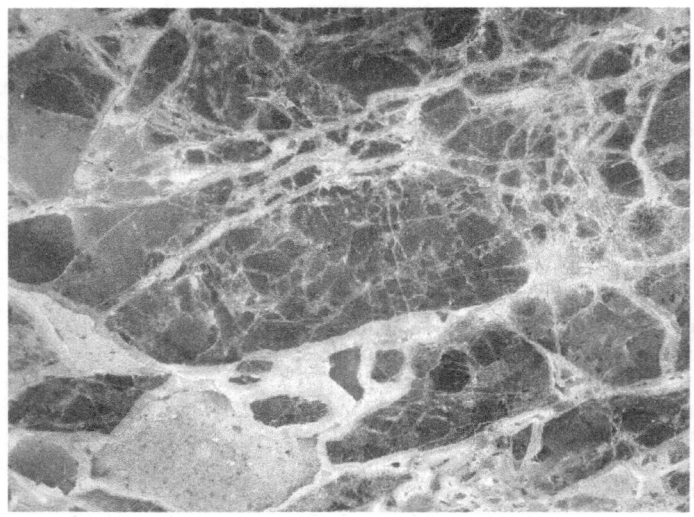

Peridotite, dredged from the depths of the mantle

All igneous rocks apart from pyroclastic debris are molten at some point, but the magma itself evolves to create different types of rock. The heavy mineral olivine crystallizes early and sinks to the bottom of the magma chamber. So does the heavy mineral chromite, the ore for chromium, which later appears in the ground as black, heavy bands. After the magma has almost completely crystallised, there remains a mixture containing the most volatile minerals, chlorides and fluorides, and metallic elements in solution. This find fissures in the surrounding country rocks, forming dykes which cool to yield silver, copper and zinc, amongst other things.

Sedimentary rocks are formed by deposition – hence their characteristic stratification – and again have important subdivisions. First are the clastic rocks, formed from particles of other rocks. The term *clastic* derives from *clast*, or a piece of older weathered rock. The clastics include sandstone, gritstone, clay, mudstone, siltstone, marl (a calcareous clay) and shale. (Note that the term "mudstone" appears on modern geological maps to indicate any rock formed from mud, including shale. It is formed from the finest sediments, whereas siltstone is made of slightly heavier fractions.) Clay itself is composed of tiny particles of the silicates of aluminum, sodium, potassium and magnesium, the worn-down fragments of igneous and other rocks (and of the feldspar minerals within them).

Sandstones made from pure quartz are comparatively rare. Types which contain a lot of feldspar as well are known as arkose. One famous landmark composed of arkose is Ayer's rock or Uluru in Australia, late Precambrian in age.

One type of sedimentary rock which is widespread in the older geological regions of Britain is the greywacke, often occurring in deposits thousands of feet thick, as in the Lake District and the Southern Uplands of Scotland. It is a type of coarse sandstone with ill-sorted fragments of varying sizes. Many of these formations are thought to have been created by fast-flowing underwater currents called turbidity currents, and so are called turbidites. In the field, however, there is often a clear difference between greywackes and turbidites, because greywackes present an appearance of an unconsolidated grey mass, whereas turbidites show a rapid alternation of bands, coarse sandstone layers being separated by fine-grained mudstone layers.

These types of clastic rocks often show one form of bedding or another, whereby their outcrops can be seen to be divided into different layers, the result of different phases of sedimentation. Minerals such as white mica often accumulate on the bedding planes of sandstones. The presence of such bedding planes makes the rocks easy to split, even if they are very hard rocks. They are also invariably the best place to find fossils. Clays show a finer form of stratification called lamination, where the layers are less than a millimeter thick. These can make the rock appear layered in outcrop. One well-known type of clay in England, the Blue Lias, obtains its name from a corruption of the word "layers" into "Lias". Note however that bedding is not confined to sedimentary rocks, as it can also be found in volcanic lava flows and ash deposits.

Another clastic group is made up composites of other rocks – conglomerate (rounded pebbles within a finer matrix, sometimes representing ancient beach material); breccias (angular pebbles also within a finer matrix); and boulder clay, a clay matrix including all manner of stones and pebbles accumulated by ice sheets and glaciers. Another rock in this group is olistostrome, a word formed from the Greek *olistomai* (to slide) and *stroma* (accumulation). These rocks are made up of a chaotic mass of heterogeneous material accumulated under the sea as a result of the gravity sliding or slumping of unconsolidated sediments. Modern examples can be seen in the Bay of Bengal. These rocks all lack bedding.

The second sedimentary group is the non-clastic rocks, formed by the deposition of minerals or organic matter – for example rock salt, iron ore, flint and coal. The third main group is the calcareous rocks, composed mainly of the mineral calcium carbonate and frequently, though not necessarily, formed from organic remains – limestone and chalk. A variant is the rock which forms from magnesium carbonates, known as magnesian limestone or dolomite. This is thought to occur where magnesium-rich fluids have altered the chemical structure of rocks which were originally laid down as limestone. (Note the similarity of names, but not of rocks : dolomite is magnesian limestone, but dolerite is a fined-grained igneous rock normally injected into existing rocks as dykes or sills.) Precambrian (very old) limestones formed from purely chemical processes, but only in warm seas. Cold water retains its carbonates, but they can be released in tropical waters – as happens today on "carbonate platforms" around Indonesia or beneath the Great Barrier Reef.

Often associated with chalk and limestone is flint. This is a form of chert, a siliceous sedimentary rock derived from the fossil remains of tiny sea creatures called radiolarians, which construct their shells from silica. One ornamental variety is known as jasper, commonly coloured red by iron staining. This can be polished and used to make vases and boxes.

Of the three main rock groups, only the sedimentary rocks are fossil-bearing.

The metamorphic rocks are formed from either of the above groups, by the action of heat and pressure normally at great depths beneath the earth's surface. Sandstone turns into quartzite, clay into slate, limestone to marble. Further heat and pressure on slate causes new minerals to form, notably the greenish chlorite, to produce a rock known as phyllite. Higher temperatures and pressure still, plus perhaps the passage of water through the rock, bring about the appearance of shiny black and white mica and red garnets (or andalusite). The rock is by now coarse-grained, and called garnet mica schist. One step up from this, other minerals appear, notably blue kyanite. After this point the heat and pressure cause the formation of banded gneiss, and then the rock finally melts to produce migmatite (a mixture of unmelted gneiss and granite which was once molten) and finally granite itself. All these higher grades of metamorphic rock are associated with the roots of ancient mountain ranges, and feature strongly in places such as the Grampian Mountains of Scotland where the different grades of metamorphism are displayed regionally.

Lying as it does near the end of the line for rock transformation, gneiss is a very common rock on ancient continental shields which form the bedrock geology, surface or subsurface, of most of the land on earth. On geological maps, the terms metapsammite (derived from sandstone) and metapelite (derived from clay) are found to describe metamorphic rocks, usually very ancient ones.

Marble is created from carbonate rocks which have been completely recrystallised, with colours supplied by non-carbonate impurities. This is unquestionably the most beautiful of all rocks, sought after down the ages for statues and the most expensive of buildings. The most famous marble in the world comes from Carara in Italy, but there is an equivalent in the United States, the Yule marble of Colorado, used to fashion the tomb of the Unknown Soldier at Arlington National Cemetery, and the Lincoln Memorial in Washington. Again, there is a tendency amongst builders to refer to ordinary limestones which can take a polish, especially Purbeck, as marble, but this is not metamorphosed, and so not marble.

Quartzite is known as one of the toughest of all rocks which is so hard that it barely yields a soil. Outcrops frequently remain where other rocks have eroded away. Note examples are Schiehallion in Perth and Kinross, Scotland, a famous pyramidal peak, and Stiperstones in Shropshire, which has weathered into fantastic shapes such as the Devil's Chair

Although it has always been known that metamorphic rocks are produced at high temperatures and pressures, nowadays it is possible to be more specific, to know exactly what temperature and pressure produced a given rock. A whole range of presses and other equipment, some of it very high-tech – such as a "multi-anvil device" – can be used to create the conditions found well down into the mantle. These machines can even be used to create synthetic diamonds.

The heat and pressure of metamorphism leave characteristic marker minerals. Three of these are forms of aluminum silicate, each with the same chemical composition, but each with a different crystallography. There is a well-known diagram which shows the relationship between them. Very high pressure produces kyanite; very high temperature, sillimanite; and lesser extremes, andalusite.

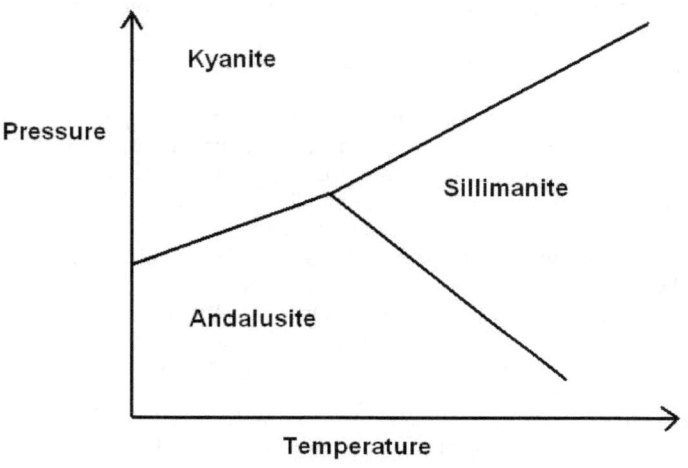

Especially in regions of old, Palaeozoic geology, such as the English Lake District, it can be surprisingly difficult to identify rocks and minerals, especially when they are dark and fine-grained. The geologist or the mineralogist has a whole battery of techniques, some of which cannot be deployed in the field. The first things to look at are the texture (fine, medium or large-grained) and the colour. The colour can be deceptive in field specimens, and is obtained by reducing a specimen to powdered form, or rubbing it on a piece of unglazed porcelain (for this reason the colour is known as the streak). One well-known example is haematite, which is often dark brown or black in the field, but invariably Indian-red when powdered – indeed the North American Indians used to paint their faces with it. Then there is lustre, which may indicate the presence of a metal, and hardness. For the latter there is a scale, called the Moh's scale, used for individual minerals, and sliding from 10 (diamond) to 1 (talc). Other attributes include weight or specific gravity, cleavage, magnetism, susceptibility to acid attack (indicator of carbonates, notably limestones), electrical conductivity (for minerals), general appearance ("habit") and a variety of blowpipe tests such as flame colours, and even X-rays. A definitive answer may only be obtained by examining a thin slice of a specimen under a microscope, a process which should show individual minerals.

For rocks alone, the presence of bedding planes is crucial as these rule out igneous and metamorphic rocks (which can still, however, show cleavage). The fact remains that some rocks are difficult to identify, especially fine-grained ones, which in my experience often turn out to be tuff!

****

As part of my degree at Oxford, I was required to complete a dissertation about a geographical area of about 500 square kilometres (150 square miles). I thought I would like to specialize in the effects of Ice Age glaciation and there was an area on the Scottish coast that looked just right for this purpose. This is called the Morvern peninsula and it is on the mainland, just opposite the Isle of Mull. So it came about that I packed up a tent and, with very little money in my pocket, set off for Morvern. The rules stated that I was supposed to spend a minimum of two weeks there.

I took the train to Glasgow from my home in Halifax, Yorkshire and then hitch–hiked past Loch Lomond and across Rannoch Moor, arriving in Oban where I slept in the open in my sleeping bag. I then caught the ferry to Lochaline, the only small town on Morvern. Nobody else got off there. To my dismay, there were no proper shops. I only found one small store which didn't have much at all, but it did sell alcohol. I remember buying a large packet of oatmeal biscuits, four cans of Guinness and half a bottle of whisky.

I set off to walk along the only road across the peninsula, but soon stepped off this, hoping to poke about a bit and find some interesting glacial moraines. However, as soon as I left the road, I was deep in a bog. This was not going to work.

At the end of my first day I had got quite a way across Morvern. There is almost nothing there – a sand quarry at Lochaline; three or four small crofts on the south–west coast; a Forestry Commission plantation at Fiunary; and otherwise, it is all moorland, given over to sheep, grouse and deer. There are hills rising up to nearly six hundred metres (two thousand feet), sweeping down into pretty lochs, such as Loch Arienas, but in terms of the Scottish Highlands, the scenery is not spectacular. In the east there is an area called Kingairloch where there is a castle and some hunting lodges for the use of the huntin', shootin' and fishin' aristocracy after the Glorious 12th August every year, when

Scotland suddenly fills up with these people. This was outside my study area and I didn't go there.

I pitched my tent and sat miserably eating the oatmeal biscuits and drinking the Guinness, which at least made my load lighter the following day. In the night it rained quite heavily. The skies were gloomy as I resumed my journey, and it was misty on higher ground, occasionally drizzling. I had no proper weather protection and risked getting seriously wet, with no means to get dry again. By the end of the second day supplies were running out. This fieldwork was supposed to take two weeks, but it was so dismal that I cut and ran for home.

I cut inland and hitch–hiked back to Glasgow, sleeping one night in a partially constructed house. To my surprise, I blundered into someone else sleeping rough in this same house. I offered him a swig of my whisky as a gesture of friendliness, which he did not refuse! It was a Sunday when I finally made it back to the Lake District, where I caught a bus to Keighley, in Yorkshire. As I passed through Skipton and the Dales, the only thing left in my bag to eat or drink was the half–bottle of whisky, so I thought, well, I may as well drink this. By the time the bus arrived in Keighley, I was pretty cheerful.

However, as I walked through the bus station, I had to pass three or four skinheads, loitering on the pavement. My hair was long in those days, and skinheads did not like people with long hair. They growled as I passed, but I made it onto my bus. When the bus set off, it went past these same skinheads, and emboldened by the whisky, I flashed them a V–sign. The effect was instant. They shook their fists and howled – then the bus stopped. For one heart–stopping moment, I thought they were going to catch it, but it set off again; that could have been very unpleasant. I never drank whisky on a bus again.

When I got back to Oxford I discovered that by coincidence, someone else had written a dissertation about Morvern only the previous year, and worse, it had won the prize for the best dissertation! I met the girl who had written this, who was now at Cambridge. She was a slightly horsey, aristocratic girl and she clearly knew a lot more about Morvern than I did. It appeared that she was able to spend whole summers there.

"So how were you able to spend so much time in Morvern?" I asked.

"I have two sets of relatives who have hunting lodges at Kingairloch. I've spent a large part of my summers there for several years."

I had been expecting this, she looked the type. She subsequently wrote me several amazingly long letters on the subject, telling me that it was best to keep off the moors in the deer stalking season and that kind of thing. I did rather wonder if there was a hidden agenda.

Chapter 3 – Modern Geology

The world of the geologists was to be turned upside down as a result of the development of paleomagnetism and the mapping of the ocean floor from the 1950s. Only then was the mechanism of plate tectonics revealed for the first time, and this mechanism must be understood before consideration of any ancient sediments. To look at the globe today, it may seem blindingly obvious that the African plate, spearheaded by the Italian peninsula, is ploughing into the European plate, forcing up the Alps straight ahead, and crumpling the landscapes of Spain in the west and Turkey in the east as it does so. The same can be said of the Indian plate, pushing under the Asian plate and throwing up the Himalayas and the Tibetan plateau as it forces its way underneath them. Meanwhile it had long been remarked that the east coast of South America fits remarkably snugly into the "armpit" of West Africa. Again, Madagascar can be made to fit very closely to the coast of East Africa. Matches such as these work even better when the edge of the continental shelf, rather than the edge of the current coastline, is used in the fit. (The continental shelf lies mostly at less than 200 m/650 feet below sea level. Below this depth the sea level plunges rapidly to the 3-5 km/2-3 mile depths of the abyssal plain.)

It had also long been known that the similarity of fossils in South Africa, South America and Australia, notably of the seed fern *Glossopteris*, indicated a common history and implied the existence of an early super-continent known as Gondwana. There was yet another strand of evidence which had emerged from geological field mapping of colonial areas in the later part of the nineteenth century – the presence of glacial tills (indurated boulder clays) from an ancient ice age (in fact Permo-Carboniferous), also in Gondwana. This implied that Gondwana was likely to have been over, or at least near, the south pole, nowhere near most of it today. It was also noted that there is much similarity in the Precambrian sequences of Central Africa, Madagascar, southern India, Brazil and Australia.

The mapping of the ocean floor had begun as early as 1872-6, when a Royal Navy ship, the *HMS Challenger*, set out to collect thousands of soundings. Although the overall pattern of the ocean floor did not emerge, the expedition found two significant things. The first of these was the existence of deep sea trenches. In fact the ship stumbled upon the deepest of them all, finding the Challenger Deep in the Marianas Trench at 11,138 metres (36,200 feet). The other discovery was the Mid-Atlantic Ridge. Sounding indicated a rise in the sea floor by as much as 4,600 metres (15,000 feet) above the abyssal (ocean-bottom) plain in the middle of the Atlantic.

Mountain building on a grand scale, **Indus Valley**, Himalayas

In 1912 there was a conceptual development. A German meteorologist called Alfred Wegener, having studied what evidence there was, proposed the idea of "continental drift" whereby the continents floated apart on a bed of oceanic crust. However there was much uncertainty about the mechanism by which this could take place. (Wegener died in 1930 aged only 50 on a meteorological expedition on the ice of Greenland, his theory still unproven.)

The first hard scientific evidence for continental drift came in the 1950s from the study of the residual record of the earth's magnetic field stored in rocks, a subject known as paleomagnetism. (The prefix "palaeo" or "paleo" is common in geology and is taken from the Greek word for "old".) Certain minerals containing iron deposited in rocks lock in a record of the direction and intensity of the magnetic field when they form. Most notably, this phenomenon can be observed in basalt, the volcanic rock which underlies the ocean. When it cools past a certain critical temperature known as the Curie point, the iron mineral magnetite within it records the magnetic orientation of the earth frozen at that time. A sensitive device called a magnetometer was invented in 1956, which could measure this magnetism. Reading from rocks in widely separated regions only made sense – that is, pointed to the same north – if the continents were moved from their current positions to the places on the globe where they must have been when the rocks were laid down.

To the strange story coming from paleomagnetism came another strand of evidence, this time from the ocean floor. The invention the nuclear submarine and the opening stages of the Cold War made a much better knowledge of the contours of the ocean floor a requirement. Reliable mapping of this finally began to take place in the 1950s, when the development of sonar and seismic equipment made the physical plumbing of the depths – on occasions five miles down and more – as undertaken by the *Challenger* unnecessary. (Nevertheless, many cores of the sea bed itself have now also been recovered.) The evidence from the oceanographic surveys was collected and published in a map of the North Atlantic ocean bed produced by Bruce Heezen and Marie Tharp of Columbia University in 1957. This showed for the first time the true extent of the Mid-Atlantic Ridge, the most sensational geological discovery of the twentieth century and running the complete length of the Atlantic. It stands high above the abyssal plain, which itself is progressively older, more subsided and so deeper away from the ridge. In only one place does the ridge reach the surface, in Iceland.

The Mid-Atlantic Ridge in Iceland

Heezen and Clark realised the significance of the Mid-Atlantic Ridge immediately. For one thing, its profile closely resembled that of the East African Rift Valley, already thought to be a spreading crack in the surface of the earth. The notch at the top of the ridge was very similar, producing a pronounced rift valley – that is, a valley caused by faulting on either side of it – sometimes well over 1500 metres (one mile) deep, and 16-32 kilometres (10- 20 miles) wide .

The pair's next map of all the world's ocean floors, since widely reproduced, appeared in the *National Geographic* magazine in 1967. This map showed for the first time the existence of volcanic mid-ocean ridges running across all the world's oceans. In fact a map of these shows the world looking like a tennis ball. In any event, the new map fascinated the world. Drawn to an exaggerated vertical scale of twenty to one over the horizontal, it showed some totally unexpected features. These included massive fans of sediments carpeting the ocean floor off the coasts of India to depths of 19 kilometres (12 miles). Debris from the erosion of the Himalayas has been dumped into the Arabian Sea and the Bay of Bengal off the Indus and Ganges/Brahmaputra deltas, and has then spread out beyond the southern tip of India, on the eastern side by 3000 kilometres (2,000 miles). The undersea mechanism which can achieve this came to be known as a turbidity current. These are often triggered by earthquakes and can run at 72 kilometres per hour (45 mph), with a range far in excess of any land avalanche. They first came to attention after an earthquake on the Grand Banks off the coast of Newfoundland in 1929. Over a period of six hours, underwater telephone cables went out of operation im sequence over a distance of 600 km/350 miles, indicating a speed of 100 km/60 mph.

It was quickly understood that new ocean crust was being created along the mid-ocean ridges from within the mantle, the layer beneath the crust. Then by a process of sea-floor spreading, the new crust moved either way away from the ridge – in the case of the Atlantic, east and west. Hence South America, for example, is not drifting, but moving as part of a slab of crust or plate which is itself being newly created along its eastern border at the Mid-Atlantic Ridge. Now at last there was a theory that gave an explanation for the facts. This meant that every description of the processes of mountain building written before 1964 or so became immediately out of date.

Another part of the picture is the existence of mighty oceanic depths, the most famous of which is the Mariana Trench in the western Pacific, which is 11 kilometres (7.5 miles) deep – deeper than the highest mountain on earth (Everest is about 8 kilometres (5 miles) above sea level), but of the same order of magnitude, a fact which is not likely to be coincidental. In fact the total amplitude of the wrinkles on the surface of the earth is tiny, making it smoother than a billiard ball for its size, as they are constrained by the inward pull of the earth's gravity and the fluidity of its mantle and core. The highest known mountain in the solar system is Mons Olympus, on Mars, at nearly 22 km (14 miles). Although much smaller than earth, Mars has a solid interior which can support greater weights. Any mountain that size on earth would immediately sink into the mantle.

More evidence for plate tectonics arose from an unplanned source. During the late fifties and sixties, the Americans set up a worldwide network of seismographic stations, but the purpose of these was not to locate and measure earthquakes – it was to detect and measure the nuclear test explosions which were being conducted at that time by the Soviet Union, notably on the Arctic island of Novaya Zemlya. Nevertheless, these stations DID locate and measure whatever natural shaking of the earth's crust was taking place at the same time. This revealed a pattern of earthquake activity worldwide, but the earthquakes were not taking place just anywhere. They were overwhelmingly concentrated in certain zones, some crossing the continents, others under the middle of the Pacific. The military scientists found that they had mapped the borders of the world's tectonic plates.

As the earth is not expanding, if oceanic crust is being created along the ridges, then it is also being destroyed elsewhere. In the western Pacific it is being dragged down underneath the east Asian plates, a process known as subduction. The deep oceanic trenches indicate its downward trajectory. It dives underneath Japan, causing the tremendous amount of earthquakes and volcanic activity there as the crust melts under the earth. Again, hard scientific evidence for subduction came from deep reflection profiles produced from seismic data, which showed the very image of oceanic plates dipping into the mantle on their way to destruction beneath the continents. Note that subduction is not a feature of every continental shore – for example both sides of the Atlantic are passive margins and lack submarine trenches. Subduction is only taking place in two small places, the Puerto Rico Trench and the South Sandwich Trench.

Within the Pacific basin, the East Pacific Rise is the eastern Pacific equivalent of the Mid-Atlantic Ridge, but it is a much smaller-scale affair, rising only about 300 m (one thousand feet) above the surrounding sea floor, with a barely detectable notch at the top. However, it is creating new sea floor much faster than the Mid-Atlantic Ridge. The sea floor here is spreading apart at a rate of up to 22.5 cm (nine inches) a year, compared to 2.5 cm (one inch) a year in the Atlantic. The East Pacific Rise is both hotter and more volcanic. The reason for the fast movement appears to be equally fast subduction all around the Pacific.

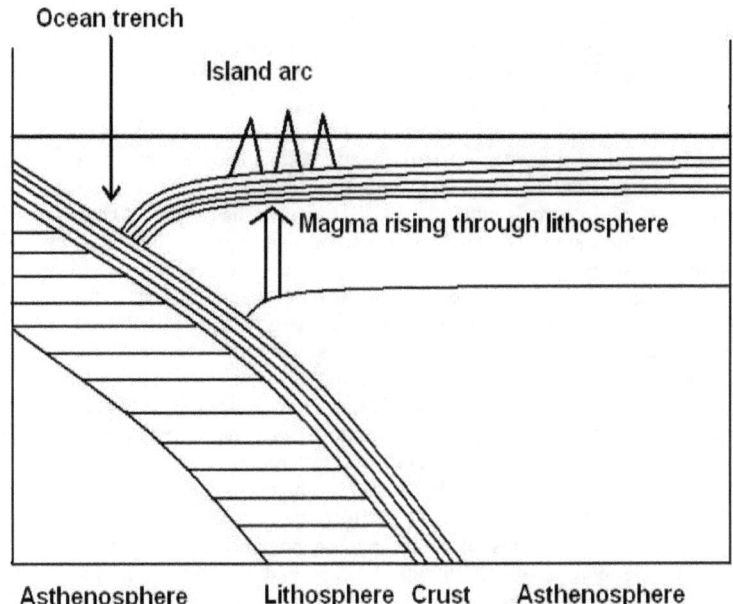

The process of subduction, where one oceanic tectonic plate dives beneath another, creating both an oceanic trench and an island arc. If a continental (rather than oceanic) tectonic plate (such as the Indian) crashes into another, the results are different – a chain of mountains, in that case the Himalayas.

****

The results of oceanic drilling show that the ocean bed is composed almost entirely of basalt, a rock which hence covers about two-thirds of the surface of the world in beds about five kilometres (3 miles) deep. Underneath the basalt is a layer of gabbro. The ocean floor lies at a depth of three to five kilometres (1.9 – 3.1 miles) and the total oceanic crust itself is about ten kilometres (6 miles) thick. These oceanic rocks have a density three times that of water. By contrast, continental rocks have an average density 2.7 times that of water. So there are fundamental differences between the oceanic and the continental crust. Not the least of these is that the continental crust has a far more complicated lithology – there are many more rocks than just basalt and gabbro. One glance at the geological map of Scotland should be enough to convince the most casual observer of that. Also the continental rocks are much thicker than the oceanic at 30-40 km (19-25 miles) – perhaps double that under the Himalayas.

Paleomagnetism then made another contribution to the emerging picture of sea-floor spreading. The magnetic record showed that from time to time, the earth's magnetic field has reversed, so that south became north, and north, south. The last reversal was about 780,000 years ago. If the theory of sea-floor spreading was correct, then these magnetic anomalies should be recorded symmetrically either side of the mid-ocean ridges. Studies by two Cambridge geophysicists reported evidence of this very phenomenon on the Carlsberg Ridge in the Indian Ocean in the journal *Nature* in 1963, ever since remembered as the "Vine and Matthews" paper. Their observations were later confirmed from ocean cores as the big, well-funded American oceanic institutes got involved (deep sea drilling time does not come cheap!) The next step was to find dates for the magnetic reversals recorded in the rocks, to see if those found around the world coincided. A new mass spectrometer was constructed at the University of California which could measure tiny amounts of radiogenic argon, and this confirmed that the magnetic reversals of 990,000 years ago – the "Jaramillo Event" – were contemporary with each other. It also allowed the rate of spreading to be measured. The history of this scientific episode was recorded in a book by W. Glen called *The Road to Jaramillo*. Paleomagnetism as applied to ancient fragments of continental rock continues to extend the history of plate tectonics back in time.

Detailed study of the oceanic ridges showed something else – they are not straight as originally thought, but are strikingly displaced, as if two planks laid side by side had a line drawn across them and one of them had then been moved along. The Canadian geologist Tuzo Wilson realised that the ridges were being moved along special faults, where no up or down movement took place. Slices of the ocean floor simply slide past each other. Wilson called these features transform faults. They are thought to develop as what is basically a linear feature, the mid-oceanic ridge, has to adjust to the curvature of the surface of the earth.

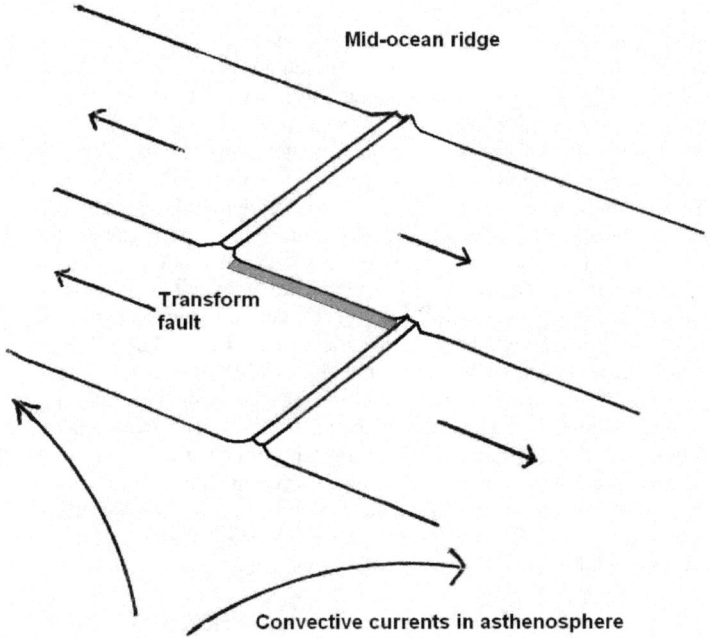

Structure of a transform fault

Fossil evidence has added to the picture of sea-floor spreading, and has the great advantage that it is much easier to examine a slide of a deep sea core than it is to book time on a mass spectrometer. The ocean floor is covered with shell debris, most of it the remains of single-celled floating animals called foraminiferans and radiolarians. There is a difference between these, as the forams build calcareous shells, and live mostly in temperate and tropical waters. Radiolarians build siliceous shells and live in cold water, along with another type of plankton, diatoms (single-celled algae, so plants, not animals). Both forams and radiolarians evolve quickly and can be used to date submarine deposits. An Eocene foraminiferan fifty million years old does not look like a modern one. The results from the cores once again confirmed the reality of sea-floor spreading – the further away from the mid-ocean ridge, the older the foraminifera which could be found in them. This evidence also showed that there is NO sea floor older than the Jurassic –

everything over about 150 million old has been subducted – so the ocean floor is much newer than most of the continents.

The confirmation of plate tectonics proved a great boost to the methods of Charles Lyell – what is happening in the earth today must have happened in a similar way in ancient periods. Puzzling features such as the mechanics of mountain folding – which seemed to require at least one slab of crust to move – were resolved. Previously mysterious formations could be re-examined. One of these is the ophiolites. These are made up of a sequence of three rocks: pillow lavas of basalt laid down on the sea floor; green serpentinite, a substance produced when material from the earth's very mantle (peridotite) is hydrated with sea water; and chert, a siliceous sedimentary rock derived from the fossil remains of radiolarians, which construct their shells from silica. This weird combination is known as the "Steinmann Trinity" after the Alpine geologist who first described it, but versions of the same thing are found in other mountain belts. The modern idea is that these rocks are chunks of the ocean floor which have ended up on land after the closure of a small ocean basin, such as is now found between island arcs (for example Japan) and the mainland. The mix of rocks in an ophiolite sequence is not, however, always the same. Some contain granitic rocks; gabbros; dolerites injected in vertical sheets; and chalk instead of chert.

(Serpentinite - also simply called serpentine, after its characteristic mineral - is the state rock of California, where it is widespread. It has been used for decorative purposes in Europe for centuries, and forms the dark bands in the walls of the cathedral at Florence, and the dark columns of the Hagia Sophia in Istanbul.)

Geologists have now learned that mountain ranges ancient and modern show themselves by a package of features – volcanic rocks with granite intrusions (known as plutons), thick folded sedimentary rocks of "geosynclinal" origin, and in certain circumstances, ophiolites.

****

Another important development during the twentieth century has been the understanding of the immense age of the earth. Even late in the nineteenth century, Lord Kelvin put it at no more the a hundred million years. The discovery of radioactivity and radiometry, the measurement of age by the process of radioactive decay of various chemical elements, changed all that. New estimates based on measurements of the relative abundance of uranium isotopes meant that by 1956, it had been realised that the earth is actually about 4.55 billion years old. This age is taken from meteorites found on the surface of the earth, thought to date from the very start of the solar system. Clearly this is an immense amount of time, enough for the surface crust of the earth to recycle and reconfigure itself again and again. The pattern of the continents 150 million years ago is unrecognizable from the continents we see today. Again, things had changed completely from the position 150 million years before that.

One man who took a central role in geochronology was the English geologist Arthur Holmes (1890-1965), eventually Professor of Geology and Edinburgh University. In 1913 Holmes published his famous booklet *The Age of the Earth*, in which he estimated the age of the earth at 1.6 billion years, at that time far older than any earlier estimates. He also championed the theory of continental drift promoted by Alfred Wegener at a time when it was deeply unfashionable with his more conservative peers. One problem with the theory lay in the mechanism of movement, and Holmes – presciently, as it turned out – proposed that Earth's mantle contained convection cells that dissipated radioactive heat and moved the crust at the surface.

His second famous book *Principles of Physical Geology* was first published in 1944, and it is for this that Holmes is remembered by students of the time, including myself. In its second edition (1964), it was a fearsome great blockbuster of a book in red covers. Extending to over 900 pages, it was chapter and verse for the subject in the 1960s – but what were these strange, speculative chapters at the end of the book, so different from the solid certainty of the rest of it, on the subject of plate tectonics? Could it be true? At that time (1968) the theory had scarcely percolated down to undergraduate level.

****

The ocean floor has continued to yield new surprises as technology developed. Associated with the mid-ocean ridges themselves were later found new, strange phenomena – black smokers. These were only detected when illuminated by the lights of the submersible *Alvin*. When sea water permeates cracks in the open, molten crust under the ocean, it becomes superheated, to 300 degrees Celsius or more, but cannot boil because of the immense pressure. It becomes charged with minerals in solution, then issues from hydrothermal (hot water) vents, building fantastic chimneys of iron sulphide (pyrites), belching black smoke, and supporting an amazing and colourful fauna of tube worms, gigantic clams and other previously unknown creatures. The base of the food chain consists of bacteria which rely on chemical energy for their metabolism rather than sunlight and photosynthesis.

In fact the science of geology is still rapidly advancing. This is now a world of electron microscopes with secondary X-ray detectors and mass spectrometers. For example the study of "Paleoclimates" – the climates of ancient times – involves the very latest available scientific techniques, measuring minute quantities of the different isotopes of potassium, beryllium and other elements laid down in ancient rocks. One aspect of the new technology is the usage of zircon (zirconium silicate) crystals to provide accurate dates for rock sequences. Zircon is a very tough mineral formed in granite, which often survives erosion to reappear in tiny quantities in later sandstone sediments, even when metamorphosed. Methods have been developed to extract these crystals, which contain a minute amount of radioactive uranium. This provides a clock to date the formation of the original granite. Dates for Precambrian shield rocks 1.7 billion years old can be obtained which are thought to be accurate to within six million years, or even less.

It is due to the development of zircon dating that the ages of the geological periods are much better defined than they used to be. Forty years ago the base of the Cambrian was put at 600 million years, a figure now reduced to 542 million years.

Zircon grains (as opposed to whole rocks) are the oldest things ever dated on the earth. Some have been found in Australia which have been dated to 4.4 billion years old – only 150 million years after the formation of the earth itself.

****

The isotopic analysis of various elements has come to play a very important part in modern geology, but what exactly is an isotope? An atom has three constituents: the positively-charged protons, at the centre; the negatively-charged electrons, "orbiting" around it; and a variable number of uncharged neutrons which form the nucleus of the atom with the protons. The number of protons and electrons in the atoms of any one element is always the same, and give the element its atomic number – for example, oxygen 8, carbon 6. The chemical properties of an element derive from this number and are not changed by the number of neutrons. The different varieties of an atom represent the variable number of neutrons. For example, oxygen has three stable isotopes because it can have 8, 10 or (very rarely) 9 neutrons (plus always 8 protons, giving oxygen-16, -18 and -17). Although they do not affect the atomic number of the atom, the isotopes affect the atomic weight. This may not be a round number at all, as it is an average of the different weights of the various isotopes.

An unstable or radioactive isotope can give off rays which transform it into another isotope of the same element, or a different element altogether. If the number of protons is affected, the result is a different element. If it is the number of neutrons, the result is a different isotope of the same element. Thus it is well-known that uranium 238 decays into lead (and eventually into helium). Its half-life – the time it takes for half of it to decay – is 4.498 billion years, which makes it suitable for geological dating. Other elements, some not normally thought of as radioactive, can be used in radioactive dating, notably thorium, potassium, lead and carbon. Of these, carbon, which has two stable isotopes, 12 and 14, has a half-life of only 5730 years, so carbon dating is much more useful to the archeologist than the geologist.

It is the isotopes of oxygen which have proved most useful in the study of ancient climates. Harold Urey, also famous for creating organic chemicals in a test tube, realised that although the two main isotopes share the same chemical properties, in one respect they are different, because the lighter isotope evaporates more easily, leaving the heavier isotope behind. In practice the ratio of the different isotopes can be used to work out the temperature at which carbonates formed on the sea floor. Although oxygen-18 only represents one atom in 500, its presence can still be measured by modern instruments. At zero degrees Celsius or a little more – the temperature of the bottom of the ocean – a significantly higher proportion of the heavier oxygen-18 is incorporated into rocks than at the warmest ocean temperatures (about 25 degrees Celsius).

Chapter 4 – Atmosphere and Ocean

One area of modern science which has contributed to geological thinking is the study of atmospheric chemistry. The chemistry of the air has a direct impact on geology because significant changes in the amount of greenhouse gases affect the climate and so the plants and animals that live on the earth. The geological column is divided into periods and epochs marked by changes in biota, many of which must have been caused by changes in the atmosphere. Each major period has a distinguishing fauna of fossils, and ends when that fauna disappears from the fossil record, five times in a mass extinction of greater or lesser extent. Great changes in life on earth must have occurred at the boundary of the geological periods. The fossil record indicates sudden shifts from one steady, long-lasting climatic state to another.

Realistically these sudden jolts can only have been brought about by changes in atmospheric chemistry and so temperature. The levels of carbon dioxide and the much more potent greenhouse gas methane are certainly implicated. This is an area of focus for current research in the universities, as the various proportions of the isotopes of carbon, for example, are indicative of atmospheric conditions when the rocks which contain them were laid down. Clearly, whatever caused the large fluctuations in the temperature of the earth which led to natural mass extinctions long before the appearance of mankind could happen again. However, there are direct means of altering the atmosphere of the earth – massive volcanic activity running too quickly for the atmosphere to adjust, and meteor impact from outer space. There is certainly one period boundary in which the impact of a large meteor is thought to have altered the climate drastically and virtually overnight, the so-called K-T boundary between the Cretaceous and the Tertiary when the dinosaurs disappeared; but this is disputed.

First, what is a greenhouse gas? It is one which admits the rays of the sun, which arrive on earth in short waves, but does not let them out again when they are radiated back out into space as long waves. The most contentious and in some ways surprising of these gases is carbon dioxide. It only absorbs outgoing radiation at wavelengths of above 12 microns and a small amount of gas present in the atmosphere – such as was present before the Industrial Revolution ever began – captures all the radiation available at those wavelengths. Increasing the concentration in experiments has NO direct effect (see Tim Flannery, page 27). Moreover, the amount of the gas present in the atmosphere has increased from 280 ppm (parts per million) in the pre-industrial age to 380 ppm today, yet the climate was as warm in the Medieval Warm Period (approximately 950-1250 AD) as it was in the period 1960-90. An increase in concentration of 35% ought to have made a difference! Even today, with the amount of the gas increasing every year, the evidence for global warming is weak and even contradictory after 1997. For example, although thre is plenty of evidence that the Arctic ice is thinning and melting, the opposite is true of the Antarctic. Here, the area of sea ice is thought to have increased by 1.523 million sq km/595,000 sq miles above the average recorded by satellites since 1979 by 2013.

Yet the new science of geochemistry tells us that whenever the earth has become a hot place in the geological past, this has been reflected by high concentrations of carbon dioxide in the atmosphere – but is this a cause, or an effect? There is strong evidence that low temperatures produce low levels of carbon dioxide, for one simple reason – cold seas are biologically much more productive than warm ones. At the base of the food chain in the cold waters are phytoplankton, which can grow into great blooms if sufficient nutrients are present, and these draw down massive amounts – billions of tonnes – of carbon dioxide from the atmosphere. So when cold water spreads, so do the phytoplankton, and atmospheric carbon dioxide falls. When warm water spreads, the opposite happens and carbon dioxide is released back into the atmosphere. This point is discussed later in the book as there is evidence for it at the time of the melting of the Pleistocene ice sheets. (The importance of phytoplankton is not to be underestimated – they are thought to contribute HALF of all atmospheric oxygen.)

The composition of the air is approximately 78% nitrogen, 20.9% oxygen, 0.9% argon, 0.038% carbon dioxide and then tiny percentages of other gases including water vapour, methane, ammonia, nitrous oxide and sulphur gases. This improbable mixture displays what scientist call a reduction in entropy, or a persistent disequilibrium, unlike anything else in the solar system. The Second Law of Thermodynamics predicts that without external inputs, systems become disorganized and are then said to possess a high degree of entropy, or state of disorganization. It is apparent that the atmosphere of the earth is highly organized. This is because this mixture is entirely the result of the organic processes taking place on earth. If there was no life, it is likely that the atmosphere would be composed of carbon dioxide, as it is on Mars. Furthermore the proportion of gases is not at all reflected in their importance to the cycle of life. Nitrogen and argon act as more or less inert fillers, bulking out the oxygen, which would become dangerous – it is highly combustible – at higher levels. Some authorities, notably James Lovelock in his Gaia thesis, maintain that the mass of biota upon the earth act to keep the earth's atmosphere in or close to its current equilibrium, so that, for example, if the amount of oxygen rises, then reducing gases (which inject hydrogen into the atmosphere) such as methane will be released, over time, to bring it down again (see below). Gaia is a system of controls and feedback mechanisms which always work to ensure that the planet maintains a comfortable environment, and especially temperature, for life.

Methane is twenty-four times as potent as carbon dioxide as a greenhouse gas, and is named as a possible factor in climate change at several points in this book. The biota of the earth produce a great deal of methane – much of it is generated by anaerobic bacteria (living with no access to oxygen), but a certain amount of it famously coming from the farting of cows. It also arises from the rotting of vegetable matter. Photosynthesis breaks down carbon dioxide (CO_2) into carbon, which is used by the plant, and oxygen, which is released. In chemical terms, the carbon dioxide is reduced to carbohydrate by the internal mechanisms of plant cells. Some of the carbon is turned into glucose to provide the plant with energy, and some of it is built into the structure of the plant, for example to make a tree trunk. When the tree dies, this carbon can be buried or "sequestered" – it is a carbon sink. This carbon can be released over the short term by the tree rotting in aerobic (oxygenated) conditions, or held over the very long term to form coal. At some point it is likely to reappear in the atmospheric cycle as methane (CH_4). However, methane is not stable in the atmosphere of the earth. It reacts with free oxygen to form carbon dioxide and water vapour.

$$CH_4 + 2O_2 = 2H_2O + CO_2$$

The only way that the release of buried methane could change the climate would be by it acting indirectly to produce large volumes of carbon dioxide or water vapour. In fact the role of methane in the atmospheric system is thought to be to keep the level of oxygen down by taking it out of the system in this way. Otherwise the free oxygen released by plants in photosynthesis would build up to dangerous levels. So in fact the atmosphere NEEDS methane, which is simply returning the carbon which the plants took from it in the first place.

The earth's processes also produce a great deal of ammonia (NH_3), in fact about 1,000 million tonnes of it a year. It is a waste product for many organisms, secreted in the form of urea or uric acid. Again it is a tiny constituent of the atmosphere only because it is unstable as it too reacts with oxygen.

$$4NH_3 + 5O_2 = 4NO + 6H_2O$$

As it releases so much reducing gas (hydrogen) into the air, its principal function is thought to be to reduced the acidity of the rain to a pH level of about 8, the optimal level for life. Without ammonia this pH level would drop close to 3, about the same acidity as vinegar. Imagine what would happen if the rain came down as vinegar. Another way to look at this is to ask, if the atmosphere is 78% nitrogen, where did it all come from? It is not like that on Mars. The answer must be that it is what is left over after the (organic) ammonia has reacted with oxygen to form water vapour.

At any event, as Lovelock points out, any outside observer analyzing the chemistry of the earth's atmosphere from a distance – as we do for Mars and Venus – would immediately conclude that the earth must contain life. Methane and ammonia must be being produced in truly vast quantities for there to be any present at all in the air, as both are so unstable in the presence of oxygen.

No one doubts these days that the atmosphere is a product of life, but to take it a further stage, Lovelock believes that it has evolved with life over billions of years, to protect life. The sun is thought to produce 30% more energy now than it did when the earth was young. At that time the atmosphere would indeed have been mainly carbon dioxide, which would have kept the earth warm. Over time the atmosphere has changed in composition, all the while maintaining similar temperatures. The idea is that the earth has managed to keep cool as the sun has become hotter by the removal of atmospheric greenhouse gases.

However, if you bought such a thermostat as Gaia, you would soon want your money back. It would be no use if it resulted in the disappearance of 90 per cent of the marine life it was supposed to keep comfortable, as happened at the end of the Permian. Again if it allowed the temperature to reach a torrid 28 degrees Celsius worldwide, then watched it plunge by half to create a highly erratic series of ice ages, as happened between the Eocene and the Pleistocene, you might conclude that the instrument was no use. Lovelock even claims that it was Gaia trying to cool the planet which led to the ice ages (page 43, *The Revenge of Gaia*). He never mentions the configuration of the continents, the real cause of the ice ages. Perhaps he should stick to chemistry, as geology is clearly not his strong point – on page 21 he twice mentions the Cretaceous period, the time of the giant dragonflies and high levels of oxygen. He means the Carboniferous, of course, which is what he calls it on page 20.

In fact Lovelock is rather in the same position as a scientist trying to understand how life got started – he is faced with the end product, so it must have happened, but how exactly? Lovelock deserves credit for explaining the organic origin of the atmosphere, but his Gaia thesis of 1978 is distinctly short of mechanisms to explain how exactly the atmosphere is controlled, and in fact, we still do not understand it.

****

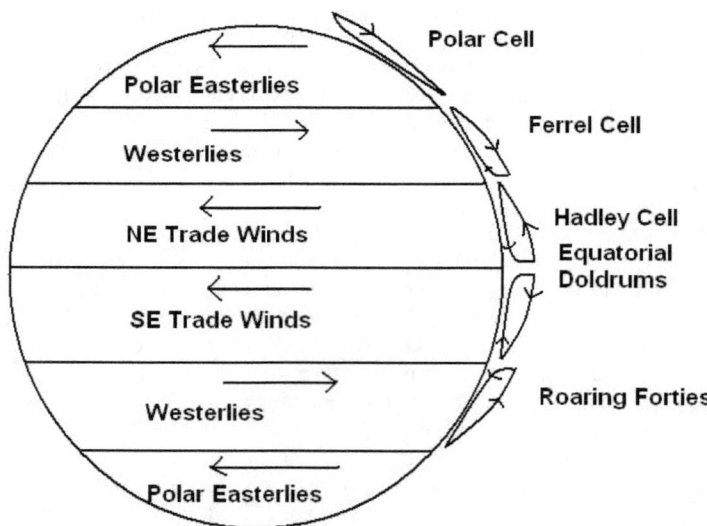

Planetary wind circulation

Deserts are an important feature of the planet. Obviously they are present on the earth now, in fact covering 20% of the land area, but they were equally important in the geological past, and the stratigraphical column is full of rocks laid down in and around them. I remember learning at school that the atmospheric circulation is responsible for the deserts of the world, and looking at the map, and thinking, that's not quite right, is it? The circulation divides into three cells, the Hadley cell from the equator to the subtropics, and then the Ferrel cell and Polar cell beyond that to the poles.

In the Hadley cell, the air rises at the equator to about a height of 17 kilometres (10.5 miles), where it meets the tropopause, the boundary between the troposphere (the lowest layer) and the warmer stratosphere above. The average rate of cooling of this air is one degree Celsius per 165 metres (536 feet). As it cannot penetrate the temperature inversion (hot air above cooler air) at the stratosphere it then moves polewards at a height of 10–15 kilometres (6-9 miles), to descend in the subtropics, then to move back towards the equator at the surface. There is low pressure along the equator, pulling air into what is called the "Inter-Tropical Convergence Zone" where air from the Hadley cells either side of the equator converges. The old sailors called the ITCZ the doldrums because of the characteristic light and erratic winds. The term "ITCZ" is often preferred to "Equatorial Convergence Zone/Low" because it can be pulled well away from the equator by regional effects, as it is in India during the monsoon. The converging air masses bring heavy rain along the ITCZ. At the subtropical end of the Hadley cell, the reverse happens. The air here is descending and warming as it does so. Warm air can hold a lot more water vapour than cool air. There is a string of high pressure cells along the subtropics. This descending, expanding ("divergent") air gives rise to the great deserts of the world. They are located at around 20 to 30 degrees north and south of the equator, known as the Horse Latitudes because that was where the old sailors used to have to eat the horses through lack of wind to move.

The surface winds blowing in towards the ITCZ are easterlies known as the Trade Winds. The air does not move in a north-south direction as might be predicted from the Hadley cell, but is deflected to the west by the Coriolis force – to the right in the northern hemisphere and to the left in the southern. On a planet which did not rotate, air would move directly from high to low pressure. Because the earth does spin, the winds instead blow parallel to the isobars (lines of equal pressure) rather than across them at right angles. So the air circulates clockwise around high pressure areas in the northern hemisphere, gradually blowing outwards (divergence), and anti-clockwise around low pressure areas in the northern hemisphere, gradually blowing inwards (convergence). When the easterly Trade Winds cross the equator, as they do in summer when the low pressure in the Gangetic Plain of India drags the ITCZ northwards, they become south-westerlies.

Around the equator itself, things are a little more complicated, because the Coriolis force has no effect here. The winds do not invariably, or even commonly, blow parallel to the isobars, hence the erratic wind patterns.

So much for the theory, but now let's look at the map! What's going on here? There are terrific deserts in North Africa, Arabia and in Australia. However at similar latitudes along the Tropic of Cancer are also found the only three hurricane or typhoon belts which exist in the world – the seas off south China (South China, East China and Philippine), the Bay of Bengal and the Caribbean! Not surprisingly there are no deserts here. The Tropic of Cancer runs through the very middle of the Sahara Desert, but also runs close to Hong Kong and Calcutta, and bisects the Straits of Florida between Florida and Cuba.

Just looking at the map, something is rather clear. Africa has a very short coastline. In all the vast area of the deserts which stretch from the Atlantic coast of Africa, through Arabia and on into Iran, Pakistan and the Thar desert of India, there are only three embayments of the sea, the Mediterranean Sea, the Persian Gulf and the Red Sea, all of which are virtually landlocked, highly saline and uninfluenced by major currents. In Australia there are simply no embayments of the sea in all the desert areas of the centre and west. The situation in some coastal desert areas is exacerbated by the upwelling of cold water offshore, notably the Humboldt current which flows off the coast of Peru and Chile where the Atacama desert is found. Most notably of all, there are virtually no offshore islands adjacent to any of these deserts.

The situation is quite different in south-east Asia where the seas off south China are surrounded by either the mainland or by islands, large and small. The same applies in the Caribbean. The areas which generate the tropical storms have many island groups – the West Indies, the Philippines, Indonesia, the Andamans. It is apparent that though surrounded by warm seas, the air over North Africa and Arabia must be very stable. It is equally apparent that this cannot be the case over the South China seas, the Caribbean or the Bay of Bengal, which are at the same latitudes. The cause must be the presence of island archipelagoes, which nudge the Trade Winds, setting off vortices which grow into hurricanes. A meteorologist would say that the islands increase the vorticity of the atmosphere. They also create very active seas, with strong tidal currents flowing between them – the same tide as on an open coast must surge through the constrictions they create. The layout of the continental land also tends to act as a funnel to incoming, wet tropical maritime air, notably in the Bay of Bengal. Because of the low-lying and densely-populated coastline, the human cost of hurricanes is higher here than anywhere else in the world.

A hurricane depends for its energy on a continuous supply of warm water (vapour), from which large amounts of latent heat are freed as the water vapour condenses. No hurricane can form where the sea surface temperature falls below 27 degrees Celsius. This is why these storms always die down out over land – but die down is perhaps the wrong word, as they can bring torrential rainfall many miles inland. There are two further conditions which affect the location of hurricanes. One is that a high-level anticyclone must lie above the low-level cyclone if it is to develop into a hurricane, to draw off the ascending air. Such anticyclones are rare along the equator. Secondly hurricanes will not in any case form within five degrees either side of the equator, because the Coriolis force is negligible in this area and so cannot impart the spin characteristic of a hurricane.

Desert scenery from the **Wadi Rum**, Jordan – a natural arch cut in soft sandstone

So the reason why North Africa, Arabia and also Australia have such large deserts is that they lie beneath areas of atmospheric subsidence, but also because they lack exposure to warm, unstable seas, and they lack offshore archipelagoes. It is noticeable, however, that it pays to be on the eastern side of the continents in these circumstances, because the Trade Winds are easterlies, so blow straight into the eastern coast of Australia, for example. The east coast of Australia is not a desert. The eastern part of Argentina is notably wetter than the sub-Andean west and south, which are deserts. The area on the far side of the Andes in the Horse Latitudes, the Atacama desert, has been dealt one of the poorest hands of cards in the world. It is in an area of the subsiding Hadley cell air; it is completely rain-shadowed from the prevailing easterlies by the high Andes; it has no penetrating seas, or offshore island arcs; and it lies next to an area of cold water upwelling in the Pacific ocean. As a result it is the driest place on earth.

****

As the atmosphere is maintained in a state of permanent disequilibrium, so is the salinity of the oceans. The weight of salts in the sea seems to have remained fairly constant through geological time at 3.4% of the total, yet salts are being added to the sea all the time by erosion from the land and by the action of black smokers at the bottom of the ocean. James Lovelock calculates that, if left to inorganic devices, the level of salts would be 13%, which would be fatal to marine life.

A certain amount of salt is removed by evaporation, as takes place in tidal bays off the Persian Gulf, but this amount is variable through time and depends on the configuration of the coastline in hot places. It seems that there must be other, biological processes at work to keep salinity level steady. Some believe that the answer is likely to lie in the sequestration cycle of marine life, whereby the shells of sea creatures fall to the sea floor, taking the salts they contain with them. However as these shells consist mainly of either calcium carbonate (in the form of calcite) or silica, this seems unlikely!

Not all seas and oceans are equally salty. The North Sea and the Baltic Sea receive large amounts of river drainage and are much less salty than the open oceans; the Mediterranean, which experiences high levels of evaporation, much more so.

A salt is an electrically neutral compound of an acid and a base. Ninety per cent of the salt in the sea is sodium chloride. In fact the two dissociate and float as separated ions in sea water. However, unlike carbonates, these chemicals do not precipitate out as deposits (limestone) in warm water. There is no such thing as a chloride rock, apart from salt itself, and this forms when the water itself evaporates, forming evaporites such as are found in large quantities under the county of Cheshire. A certain amount of chlorine is taken out of the ocean as the gas methyl chloride, which acts as one of the natural controls on the ozone layer. The fact is, however, that no one really understands how the salinity of the sea is kept so low, and so constant.

The prevailing climate greatly affects the type of geological deposition which takes place in any given locality. It is clear from paleogeography that the constituent parts of the continental crust have moved about, and that the configuration of the continents has affected the climate of the whole world. One reason for this is that the most powerful influence on climate, after the sun itself, is the ocean and its currents. This system is known as the thermohaline circulation, as it is driven primarily by heat and by salt density. A scientist as early as Benjamin Thompson, Count Rumford (1752-1814) knew that water at only 11 degrees Celsius (in fact the real figure was lower) was present no more than a mile (1500 m) below the warm tropical seas off West Africa, and he drew the obvious conclusion – local water could not be that cold, so it must have come from the poles. It does.

The best-known ocean current in the world is the Gulf Stream, which flows from the Caribbean into the North Atlantic until it reaches the latitude of Greenland. It brings vast amounts of heat to western Europe – about a third as much as the Sun itself. It then sinks, as condensation at the surface increases its salinity and so weight relative to the surrounding water. The same thing happens in the South Atlantic, off the Weddell Sea in Antarctica. It is as if the Atlantic has a plughole at either end. Once at abyssal levels the water turns back towards the equator, moving very slowly at these depths. The deep equivalent of the Gulf Stream is the North Atlantic Deep Water which carries cold water eventually into the north Pacific, taking a thousand years to do so!

The Gulf Stream itself is classified as a "Western boundary current", relatively narrow and fast-flowing as it works its way up the western edge of the North Atlantic. The equivalent in the Pacific is the Kuroshio Current, flowing along the coast of Japan and Siberia. In the Indian Ocean is another such current, the Agulhas, flowing southwards along the east coast of southern Africa. These currents flow quickly along these western boundaries for reasons to do with the physics of the spinning earth – to conserve their angular momentum. Currents on the eastern side of the ocean basins are much more sluggish.

The thermohaline circulation stretches around the globe at both surface and deep levels. The ocean itself is divided vertically into different temperature levels. Even in the tropics, as noted, and according to modern measurements, the warm water forms a layer not much more than 300 metres (1000 feet) deep. Then comes a break called a thermocline, below which the water is barely above 4 degrees Celsius. Above and below the thermocline the water can and does flow in different directions. The surface currents are driven by the winds but in any case the waters will attempt to even out imbalances in planetary heat according to the principles of the Second Law of Dynamics. The practical effect of this law is to cause a hot substance in a cool place to lose heat to its environment until it is in heat balance with it, rather than to get any hotter.

The point of this discussion of ocean currents is that the above pattern is only the modern one, and if it breaks down, the geological consequences can be profound. Later in this book we will see some of the results of that from previous world geographies. For example at the time of the Permo-Triassic, 250 million years ago, all the continents fused into one, greatly reducing the ability of ocean currents to moderate inland temperatures, and much of the world turned into a desert. In an ideal world, the ocean currents would flow freely around separated continents, and the climate would be warm and pleasant worldwide, as it was in the Eocene, 55 million years ago. The gradual breakdown of the balmy Eocene climate is attributable to the deflection of ocean currents. At the south pole, a whole continent moved to block their way. In the subtropics, the gap between North and South America closed. The Bering Strait between Alaska and Russia also closed. An ice age was to follow.

There is another aspect of oceanography of importance to geology. The top layer of oceanic water where the surface temperature is above 10 degrees Celsius forms a layer which receives most of the solar insolation – the heat from the sun. Unless stirred up by hurricanes, this layer does not mix with the deeper, cooler layers and so very quickly runs out of nutrients. Hence the clear blue surface waters of the tropical seas are clear for a reason – they contain little life. Eighty percent of the oceans of the world fall into this category. Tropical oceanic water is only rich in life if supported by cold-water upwelling, as it is in the Humboldt Current off Peru, for example. By contrast, where the surface temperature of the ocean remains below 10 degrees Celsius, the top water mixes with the lower layers, ensuring that a good supply of nutrients is maintained. Also oxygen is much more soluble in cold water than warm allowing shelly plankton and krill to thrive. From the point of view of planktonic growth, cold, sub-polar water is the place to be. The same applies to the whole food chain – right up to the Icelandic cod and the blue whale – which depends on the plankton and the krill. The great natural fisheries of the world – the North Sea/Icelandic, the Grand Banks, the Achovetta in the Humboldt Current, the Pacific salmon – are all in cold water. A warm sea like the Mediterranean is a poor fishery.

This is important when it comes to consideration of the recent ice ages, when the amount of carbon dioxide in the atmosphere fell and then rose with the formation and the later melting of the ice sheets. The fact is that plankton, which draw carbon dioxide down from the atmosphere, like it out there in the cold sub-polar seas. It is known that great blooms of plankton lived and died during times of high glaciation, sequestering carbon from the atmosphere as they did so, and so keeping the planet cold.

One other aspect of oceanic water is the fact that it expands when it is warmer than four degrees Celsius. The geological record is punctured with marine transgressions – invasions of the sea on to the land – and marine regressions. In some cases the causes remain obscure. However a warming planet will certainly favour transgressions, as could be argued for example for the Cretaceous. Conversely, a cooling planet will favour regression.

Chapter 5 – Vulcanism and Faulting

If a geological enthusiast wants to take a holiday, then there is no better place to go than that marvel of geology, Iceland. In the picture in my mind, I imagine that it is about the size of Wales, but this is incorrect – it is much larger than that, nearly 103,000 square kilometres (40,000 square miles). Compare England, 130/50 thousand sqkm/sqm or New York State, 141/54 thousand sqkm/sqm. The oldest parts in the far east and west date from the Miocene epoch and are 16 million years old, but in the south, and off the south coast, the island is still forming. Surtsey was a new island born off the south coast in 1963. As we all know (some to our cost), volcanoes erupt on Iceland all the time, generally in a rather benign, basaltic kind of way. There are no volcanoes in the older, now stable parts.

On Iceland the North American plate lies on the west, and the Eurasian on the east; their meeting in the middle of the island is marked by a notch, found all the way along the Mid-Atlantic Ridge. I've stood in the notch myself – they used to hold the old outdoor parliament there, at Thingvellir, and I must say, it seems a dodgy place for a parliament.

Erosion sets in the moment that new land is created, and Iceland has some spectacular cliffs, even now under assault from the sea. However the coast is green and attractive. Inland, the vegetation disappears and towards the many icecaps, the land has the appearance of a polar desert, devoid of all vegetation. Geothermal springs are used to provide heat and power, and there is a large aluminum plant sited in the country to take advantage of the cheap electricity. There are also geysers here, which blow regularly to heights of sixty metres (200 feet), emitting jets of steam and supercharged water and stinking of chemicals. The name comes from a place called Geysir. Then there is the thermally-heated Blue Lagoon, genuinely cerulean blue, but beware the sharp rocks!

If Iceland is a very volcanic place, it is also a very hydrological one. The island sits beneath a semi-permanent weather feature called the Icelandic Low, the counterpart of the Azores High in the mid-Atlantic. The jet stream flows between these two, and a constant series of storms is generated causing heavy rain and snow over Iceland. Also a lot of the country is quite high, forcing up the clouds – the highest peak is 2110 metres (6857 feet) and there are a number of others over 1500 metres (5000 feet). Again, there is no shortage of nuclei around which water vapour droplets can condense. Once the precipitation hits the ground, there is little evaporation even in summer. In winter the water is held back in ice caps which then release torrents of water in the summer. The structure of the land is immature and has yet to be worn down to a smooth profile. Hence Iceland is famous for its waterfalls. Dettifoss in the north is the most powerful waterfall in Europe, but more accessible and so more famous is Gullfoss in the south.

Gullfoss in August

Iceland is too bleak to support many trees. In fact there is a joke there – what do you do if you get lost in the forest? You stand up. The aspect of the little capital, Reykjavik, reflects this bleakness. Nearly all the population lives here and in other coastal communities. This population has expanded sixfold in two hundred years, and now numbers over 300,000 – still not much for an island four-fifths of the size of England. Of course Iceland has undergone the most terrific economic bust in the last few years, and most visitors would say, about time too, having been obliged to pay £50 a head for a meagre meal. For some reason the skippers of fishing boats decided it would be a good idea to become merchant bankers instead.

****

Volcanoes are generally found where tectonic plates are diverging or converging, but the result is two different types of volcano. At the mid-oceanic ridges, as in Iceland, the volcanoes which result from the divergent tectonic plates produce basic rock – basalt – and are not generally highly explosive. The Pacific Ring of Fire (see below) has examples of volcanoes caused by convergent oceanic and continental tectonic plates moving together. The most common extrusive rock in these cases is the intermediate rock andesite (which is very common in the Andes), or else an acidic rock, rhyolite, the fine-grained equivalent of granite, often produced along with masses of ash after violent explosions. (Note that the consolidated form of volcanic ash is known as tuff. This is quite different from the calcareous deposit known as tufa.)

By contrast, volcanoes are usually not created where two tectonic plates slide past one another. Volcanoes can also form where there is stretching and thinning of the earth's crust in the interiors of plates, for example in the East African Rift Valley.

There is volcanic activity all round the Pacific – in Indonesia, the Philippines, the Aleutian Islands, the western United States, Chile, New Zealand and so on – the "Pacific Ring of Fire" – which is caused by plate tectonic movements. Subduction zones and their accompanying deep ocean trenches are located all around the east, west and north of the Pacific, some apparently in the middle of the ocean. A characteristic feature of the area is the island arc. These are widespread on the face of the earth, and include:

    New Zealand and Tonga
    Melanesia (New Guinea, Solomons, Bismarks, New Caledonia, Vanuatu, Fiji)
    Indonesia
    The Philippines
    Formosa and western Japan
    The Marianas and eastern Japan
    The Kurile Islands and Kamchatcka
    The Aleutians
    The West Indies

The Western Antarctic Peninsula and islands

All but two border the Pacific. These island arcs all bulge outwards towards the deep ocean. They are invariably associated with deep-sea trenches which demonstrate large negative gravity anomalies. They are the sites of active vulcanism, which is characteristically andesitic. There is usually noticed a lateral variation in the composition of volcanic rocks from andesite on the continental side to oceanic basalt on the oceanic side – the line separating the two is called the andesite line. Associated with the vulcanism is active seismicity. The origin of the earthquakes lies on a plane dipping at roughly 45 degrees beneath the island arc where an oceanic plate is plunging beneath a continental one. Despite the fact that the island arcs are oceanic, they are sitting on the edge of lightweight continental crust under which the heavier oceanic crust is dipping. The lithology of the arcs shows no sign of the Precambrian basement rocks which underlie the true continents.

When one area of continental crust collides with another, as India has with Asia, the result is mountain building, as the continental crust is far too light to sink. When oceanic and continental plates collide, on the contrary, the oceanic one is subducted, and this will certainly be signified by a string of volcanoes, but there is not necessarily a new mountain range. In two notable cases, however, there are mountain ranges where oceanic and continental crust collide – in the Andes and also in the Cascade Mountains of the north-western United States. The Cascades are the location of Mount St Helens, a volcano which erupted with a terrific bang in 1980, giving off an outpouring of dacite, an igneous lava intermediate in composition between andesite and rhyolite.

The Pacific Ring of Fire is a notorious earthquake zone, including as it does heavily populated areas in Indonesia, Japan, Mexico and Chile. An earthquake off Indonesia on Boxing Day of 2004 set of a tsunami which is thought to have killed 230,000 people in the worst natural disaster of modern times. At Tokyo, Japan, the earthquake of 1923 killed 140,000 people. Another at Kobe in Japan killed 5,000 people in 1995. Indonesia in particular if noted for some of the largest volcanic explosions ever known. The biggest is thought to have been Toba, all of 74,000 years ago, but in more recent time have been Tambora (1815) and Krakatoa (1883), which with its associated tsunamis also killed over 36,000 people.

These examples from the Pacific show that the ocean floor, though generally rather simple in its lithology, is not necessarily as simple as all that. It contains oceanic islands, sea mounts and plateaux, volcanic island arcs, fragments of continental crust known as micro continents (such as New Zealand), and even whole undersea mountain ranges.

Volcanoes occur in zones in other parts of the world. There is a line running down the Atlantic from Jan Mayen Island in the Arctic, through Iceland the Azores and the Canary Islands to the Cape Verde Islands in the south. A further belt stretches from the Mediterranean through Turkey and the Caucasus to the Himalayas.

New land formed from recent volcanic eruptions in the **Azores**

Within the Pacific, the volcanic islands of Hawaii reveal some remarkable features of the new geology which has emerged in the twentieth century. It has been known since the days of Captain Cook that these islands are volcanic, but now a pattern to the vulcanicity has emerged. The largest, most volcanic and most southerly island is Hawaii itself, otherwise known as Big Island. To the north and west lie smaller islands, Maui, Oahu (location of Honolulu and Pearl Harbour), Kauai and others, each one lower and more deeply eroded in a north-westerly direction – but this is not the whole story. Twenty miles off the south-east coast is the Lo'ihi sea mount, another volcano, hidden beneath the waves, which will one day form another island in the chain.

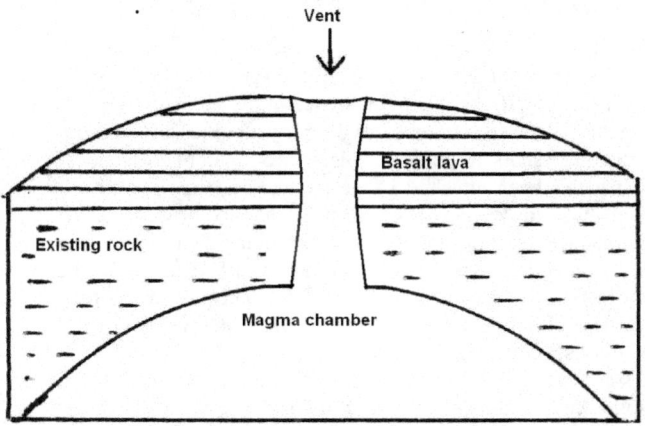

Structure of a shield volcano as seen on Hawaii (vertical scale exaggerated). Here the cone is composed of successive lava flows. In the case of explosive, Plinean-type volcanoes the cone is made of pyroclastic material.

The highest volcanoes on Big Island are Mauna Kea at 4205 metres (13,796 feet) and Mauna Loa, only 37 metres (120 feet) lower. The average depth of the ocean floor around the islands is 4877 metres (16,000 feet). Measured from the floor of the ocean, these mountains approach 9,000 metres (30,000 feet) – higher than Mount Everest. These are known as shield volcanoes, because their profile resembles the shape of a round shield laid flat. Typically the slopes are less than 8%. Because of the sheer height of these volcanoes, their bases can be very wide indeed, over 160 kilometres (100 miles) in some cases.

The terminology for different types of lava flow comes from Hawaii. First are the pillow lavas, piled up in rolls, which form under the sea. Then comes hyaloclastite, formed where the lava reacts explosively with water and air at the sea surface. Once on land the lava erupts as ropy lava or *pahoehoe*, or in a jumble of sharp-edged blocks and clinker known as *a'a*. The basalt itself is a darkish rock of basic (as opposed to acidic) chemistry, also known as mafic rock because of its high concentration of magnesium and iron. It main minerals are plagioclase feldspar, the dark mineral pyroxene, and the green mineral olivine. Also present in basalt are metallic sulphides including iron pyrites. The lava can also contain xenoliths ("foreign rocks") dredged up from deep in the mantle. Most of these are various forms of the rock peridotite.

North-west of the Hawaiian Islands, a whole chain of submerged volcanic cones has been discovered, known as the Emperor Sea-mounts. These extend all the way to the Aleutian Trench at the edge of Asia. In total the chain of volcanoes stretches over thirty degrees of latitude! Each sea mount represents an extinct volcano (or pair of volcanoes) which has first of all been attacked by aerial erosion, then sunk back into the mantle under its own weight after vulcanicity ceased. Until recently the explanation has been that the Pacific plate is moving over a lava plume, an upwelling from 3,000 km (2000 miles) down in the mantle, which has stayed in a constant position. The islands as they exist today are five million years old at the oldest, but the whole process has carried on over seventy million years.

There are some problems with the mantle plume theory. One is that there is a distinct kink in the chain, where the line of extinct volcanoes changes direction from north-west of Hawaii to north. The second is that magnetic evidence from rocks dredged up from the Emperor seamounts indicate that they formed as far north as 37 degrees – and Hawaii is at 19 degrees (north). Thirdly, seismographic evidence shows hot rocks down to 500 km/300 miles below Hawaii, but nothing deeper than that. So it may be that the lava source is not fixed, and that the chain of volcanoes represents lava erupting along cracks in the plate.

****

Another very good place to see vulcanism in action today – or for that matter to take another holiday – is the Bay of Naples in southern Italy, home to the most famous volcanic eruption of Classical times. It was here that Vesuvius exploded in 79 AD, causing a massive pyroclastic surge to bury the towns of Pompeii and Herculaneum in searing hot volcanic ash. The event was recorded for posterity by Pliny the Younger, whose uncle died in the eruption. I was given his account to translate from the Latin in my own O-level paper! Pliny was staying with his uncle at the time, and saw the whole thing from a safe distance. His uncle, Pliny the Elder, who was a commander in the Roman fleet, set off in a ship to investigate the eruption as a scientific curiosity. However his journey across the bay soon turned into a rescue mission. Having reached the shore under a bombardment of fiery pyroclasts, he found himself unable to get away due to contrary winds. He may have died from inhaling the dust as his body was later found more or less unmarked. Anyone who goes to Pompeii today can see the plaster casts of bodies which were simply engulfed in the ash in the act of trying to escape.

This type of eruption, which produces vast amounts of fragmental products but little or no lava, is known as Plinean, after the author. It is contrasted with the quiet flows of Hawaiian-type volcanoes, or the even more violent Pelean type, named after Mount Pelee in Martinique (in the Caribbean island arc). This exploded in 1902, killing all but two of the population of 28,000 of the town of St Pierre at its foot. Pelean explosions are caused by blockages of the viscous, andesitic lava in the vent to the point when the pressure becomes too great, and an explosive mass of hot, incandescent fine ash and coarse rock fragments known as *nuées ardentes* blows out. The eruption in Martinique only left a deposit of ash a few centimeters deep, and the buildings of St Pierre remained intact, but it was absolutely deadly nevertheless.

Volcanic phenomena today – this is the **Old Faithful** geyser in Yellowstone National Park

One feature of the Bay of Naples area is the Phlegrean Fields (Campi Flegrei), a mass of craters and calderas now covered by the suburbs of Naples. Charles Lyell and another famous British geologist, Roderick Murchison, visited the area together in 1828 – it was simply a "must see" for geologists. The Austrian geologist Edward Suess (1831-1914), whom we will meet later, also wrote about the area. He differed from the earlier Lyell in that he believed that while steady, slow processes did evolve the landscape, they could be interrupted by violent episodes when whole mountain chains could be thrown up. The two views were only reconciled long afterwards by the mechanics of plate tectonics. However, there is something to be said for the Suess point of view in southern Italy. For example measurements on the island of Ischia show that there has been seventy metres (250 feet) of uplift in the last 8500 years – not what we normally think of as a geological timescale.

Shop frontage carved from relatively soft **travertine**

At Pozzuoli is the Solfatara crater, which emits foul-smelling, sulphurous vapors. Around the vents are patches of rare minerals – arsenic and mercury compounded with the sulphur, with names like realgar, orpiment and cinnabar, familiar for centuries – in fact cinnabar is the principal ore of mercury, and was known to the ancients. "Solfatara" is used as a generic name for a vent emitting hot, sulphurous vapours. It is just one of a number of types of fumarole, volcanic fissures or vents which emit gases. They can be active at any stage of vulcanism but when they are the dominant form of it, they indicate declining vulcanicity. A similar form of secondary vulcanism is the hot spring. Such a spring may bring chemicals to the surface. Calcareous deposits formed in this way are called travertine or tufa. Fantastic deposits of travertine are found at Pumakkale in Turkey – I have seen them myself. They are blindingly white in the midday sun. This rock, though recently formed, is hard enough to be used in the exteriors of shop fronts.

The cause of the many volcanic phenomena of southern Italy is the movement of the Adriatic plate, vanguard of the African plate, northwards under the European plate. Italy represents the forward edge of the plate and is being twisted and deformed as a result. There are more than twenty volcanoes involved, some of them famous – Vesuvius, Stromboli, Etna (on Sicily), Vulcano itself.

In terms of geological time, another volcanic explosion in the Mediterranean will come to seem contemporary with that of Vesuvius. This was the almighty bang which blew of the top of the island of Thera, now known as Santorini, seventy kilometres (44 miles) north of Crete, in 1700 BC. It is thought that thirty cubic kilometres (7 cubic miles) of material was erupted, fifteen times more than at Vesuvius in 79 AD. This was the largest and most influential volcanic explosion in the history of mankind in Europe. Many believe that Thera was in fact Atlantis, described by Plato as a lost city beyond the Pillars of Hercules (the entrance to the Mediterranean – the only flaw in the argument). The eruption would have coated the area around the island in a thick covering of volcanic ash, destroying crops, and must have generated a massive tsunami – possibly more than one – wiping out coastal communities. It is thought to have brought the flourishing ancient civilization of Minoa to an abrupt end.

Travertine deposits at **Pumakkale** in Turkey

****

Faults are a most important feature in geology, cracks in the earth's crust along which movement has taken place. This movement can either be horizontal (lateral) or vertical. Faults in which one slab of rock moves downwards relative to another are described as normal faults. Such faults are found, for example, on either side of the Rhine Valley below Basle, a structure known as a horst (the hills on either side) and graben (the "ditch" through which the Rhine flows). The graben has slipped down between normal faults. The basic cause of these faults is a stretching of the crust. On the other hand, compression of the crust causes a reverse fault, where a slab of rock has been forced upwards. A thrust fault is a different structure, normally nearly horizontal. In mountain building, one whole mass of rock can be pushed over another at a thrust fault, a structure known as a nappe. Any geological map is likely to show dozens of faults as they are very common.

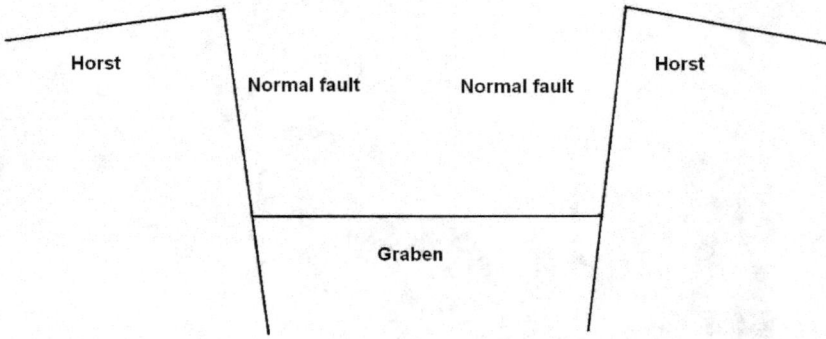

Rift valley structure, as seen in the middle Rhine below Basle

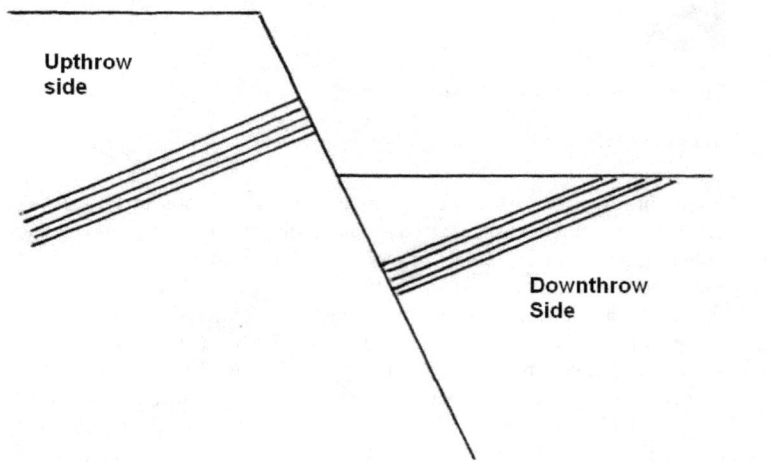

Normal fault structure.

The most famous fault in the world is a lateral one, the San Andreas Fault, which runs down the western coast of California. In fact, this long fault is associated with the East Pacific Rise. As this hits the coast of Mexico, a series of transform faults step off it at right-angles heading up the Gulf of California. The San Andreas fault is the longest and northernmost of these, heading northwards from the top of the Gulf to San Francisco and beyond.

Activity along it caused the San Francisco earthquake of 1906 which killed 3000 people and led to a movement of 6.4 metres (21feet) in some places. This fault system has been operating since the Miocene epoch 20 million years ago and shows no sign of ending. As recently as 1994 there was another severe earthquake on it at Northridge, a suburb of Los Angeles – pictures of buckled freeways were beamed around the world. The San Francisco earthquake was especially severe, ranking at 8.6 on the Richter scale. This is a logarithmic scale, in that an earthquake of magnitude 7 releases thirty times as much energy as one of magnitude 6.

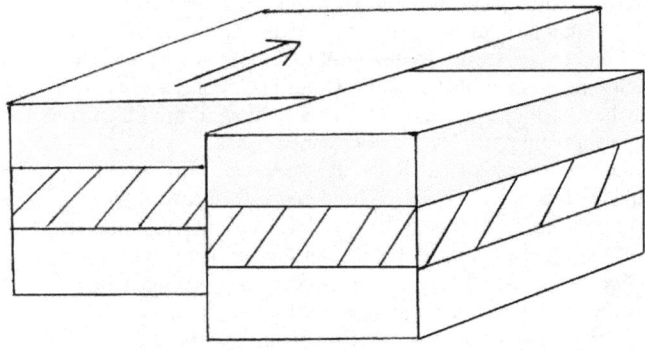

Lateral fault of the San Andreas type, where two blocks of crust or plates slide past each other without significant vertical movement

Fault have many uses to mankind as they create cracks in the earth through which molten materials containing valuable minerals can percolate from the depths. The gold which caused the Californian gold rush of the 1840s has such an origin.

One place long renowned for both vulcanism and faulting is the East African Rift Valley, considered to be the early home of mankind. This is one the geological wonders of the world. Here the earth's tectonic forces are creating new plates by splitting apart the old ones. The rifting within Africa extends northwards through the Red Sea, formed by and earlier phase of rifting, and then north again through the valley of the River Jordan. The heart of the East African Rift Valley itself is the Afar region of what is known as the Ethiopian Dome. The rifting then extends southwards into Kenya, Tanzania and Malawi – the Great Lakes region of Africa. Here the western branch contains Lakes Albert, Tanganyika and Malawi. The eastern branch that roughly bisects Kenya north-to-south on a line slightly west of Nairobi. (Note that Lake Victoria is not part of the rift valley system – it lies between the eastern and western rifts.) The topography of parts of this rift system, in practice made up of many faulted basins, can be truly stupendous. I travelled across one section of it in a small Datsun (Nissan) in 1974 and gazed in awe at the sheer task ahead of the little car as we arrived at the top of the crest, gazing down and across to the other side. However I have spent many a holiday on the seemingly peaceful and sun-kissed shores of Lake Malawi, at the southern end of the Western Rift, but even there, you had to be careful. A South African tourist was strolling along the beach one evening at the Grand Hotel in Salima, a place we often visited, when he found himself between a hippo and the Lake. It gave him a bite as it went past. He did not survive.

Associated with the East African Rift Valley are some of the world's most spectacular volcanoes, including Kilimanjaro, Ruwenzori, Mount Meru and Mount Kenya. When I flew out to my first real job in Zambia in 1973, the first landing at Nairobi in Kenya had to be called off because of low cloud. The pilot diverted to Dar es Salaam in Tanzania. However, this was a bonus because this leg of the journey afforded us the most fantastic view of Kilimanjaro, covered in snow 3000 metres (10,000 feet) below us. The mountain stands alone at 5895 metres (19,341 feet) high, and the top is frequently shrouded in cloud when viewed from below.

The most unusual volcano in the East African Rift Valley area is Oldoinyo Lengai, in Tanzania. This has a unique lava, rich in potassium and sodium carbonates rather than the usual silicates. These give great fertility to the surrounding soils and so support the mass of wildebeest which give birth to their young in this area every February, before migrating across the Serengeti.

Major faults are often the site of lakes, such as Lake Malawi. As these can be very deep, they are long-standing features of the landscape. Others in this category are Lake Baikal in Siberia and Loch Ness. Lakes which have formed in hollows created by glaciers may have a much shorter life, though, in the case of the Great Lakes of North America, which are glacial in origin, that would be "short" in a strictly geological timescale!

One very active frontier in the world of plate tectonics, volcanoes and earthquakes is the northern border of the Indian-Australian plate. This very large plate includes the whole of India, Australia, Papua New Guinea and all of New Zealand except the eastern side of South Island. The northern boundary of the plate runs under the heart of the Himalayas, thence dropping into the Bay of Bengal and heading eastwards to run south of the islands of Indonesia, including Sumatra and Java. The Indian-Australian plate is pushing north-eastwards at a rate of 67 mm/2.8 inches per year, which over the era of the white colonization of Australia and New Zealand, say 220 years, amounts to nearly 15 m/48 feet, so it is hardly surprising there is so much volcanic activity.

Along the eastern border of the Indian-Australian plate lies the New Zealand and Tonga island arc. On the eastern (Pacific) side of the islands, the Tonga and Kermadec Trenches trend south-westwards towards New Zealand. Some of the Tongan islands themselves are volcanic in origin. Within the South Island of New Zealand lies the Great Alpine Fault, a lateral (or transform plate boundary) fault like the San Andreas, the line of separation between the Indian-Australian and Pacific plates. There has been a lateral movement of 300 km/185 miles along this fault in the last few million years. As at the San Andreas, earthquakes are common along this fault, but there is no vulcanism.

Chapter 6 – The Precambrian

Probably not very long after the formation of the earth 4.55 billion years ago came the first really significant geological change – the formation of the moon. The moon is very different from the earth, as it has no molten core to create volcanic or tectonic activity, and no atmosphere to generate erosion or sustain life. The surface rocks of the moon, having survived its still-obvious bombardment by meteors, have not been subject to the recycling processes found on earth, and so are older than anything remaining on earth. Rocks brought back from the moon have been dated to 4.5 billion years old. The moon is very useful to the earth as it helps to stabilise its movements in space. In fact it is still edging away from the earth, and in earlier geological times it was appreciably nearer. The gravitational pull of the moon has caused the earth gradually to lose some of its angular momentum and so its rate of spin has slowed down. It is known from coral growth rings that there were 420 days in a year in the Lower Palaeozoic, so much more time to get things done! In still earlier times, closer to the date of its formation, there were only six hours in a day as the earth whizzed round on its own axis.

There is another body in the solar system which is also very useful to the earth, and without which complex life may never have evolved – Jupiter. This is so large that it hoovers up asteroids, meteors and comets which enter the main plane of the orbit of the planets, swallowing them up before they threaten the earth (which after all is rather a small target). This was the fate of comet Shoemaker Levy 9 in 1994, widely reported at the time.

Life is thought to have emerged in single-celled forms fairly early in the history of the earth, probably by 4 billion years ago. Charles Darwin famously imagined that this event took place "…in some warm little pond, with all sorts of ammonia and phosphoric salts – light, heat and electricity present…" However, even a single-celled organism has proved to be so immensely complex that attempts to replicate the warm little pond, and to create life from scratch, have so far failed. In 1952 an experiment conducted by Stanley Miller and Harold Urey at the University of Chicago seemed to be getting close. An atmosphere thought to resemble that of the early earth was created, containing ammonia, methane and steam. When electrical charges (test-tube lightning) were passed through it, a rich brew of organic chemicals resulted, including amino acids. These are the building blocks of proteins, which in turn are the building blocks of life. This naturally caused great excitement. Then in 1961 the Catalonian Juan Oro concocted another brew and managed to create the chemical adenine, which is one of the four "nucleic acids" (found in the nucleus) which are the building blocks of DNA, and which is also a major constituent of ATP – which provides cells with their energy.

It has subsequently been shown to be relatively easy to create these organic chemicals from a recipe in a laboratory, but quite impossible – as yet – to create proteins or DNA from them. In theory all that is needed is a chemical which can replicate itself, and a sac or cell to contain it, and then this simple construction can start to evolve. A Danish scientist called Steen Rasmussen is working on a project to make just such a cell, but his sac is made of fatty acids – and these themselves are complex carbohydrates. However it may not be as difficult as all that, as membranes called lipid vesicles can form spontaneously in nature.

For a very long period, life on earth is thought to have consisted of nothing but single-celled bacteria and another group, found in modern forms for example as red filaments in hot, chemical volcanic pools in the Yellowstone National Park, called the archaea. The most primitive forms of life known today are found in an unlikely place – around suboceanic black smokers. They are hyperthermophiles, lovers of extreme heat, surviving on hydrogen and sulphur in anaerobic environments (that is, without free oxygen). At these depths, they have no need of either oxygen or sunlight, and would have survived any surface catastrophe, including meteor strikes and snowball earth. It looks increasingly likely that these bacteria and archaea are the foundation of all other life on earth.

It is quite clear from modern examples that all known forms of life possess the same basic feature, the double structure of DNA to store the instructions and proteins to execute these instructions. All life is one, and has a single common origin. It is equally clear that even the simplest life forms are complex, and must have evolved to replace all traces of the primitive forms from which they evolved, kicking the ladder from beneath them for the emergence of any other form of life.

The early forms of life (including cyanobacteria, commonly known as blue-green algae, which are still with us) did not possess a cell nucleus – the prokaryotes. The first organisms to contain this structure are called the eukaryotes, and they too have lived on earth for a long time, 1.5 billion years or more. (Recent evidence from the Pilbara region of Australia indicates that the eukaryotes may have been on earth as long as 2.7 billion years ago.) Early eukaryotes took the form of amoebae, or were ball-like in shape, sometimes with many arms like a mine. They can be found in large numbers by dissolving shale 1 billion to 850 million years old in hydrofluoric acid. (This substance is also very good at dissolving fingernails!) The eukaryotes greatly increased the sophistication of life by including within their cells not only nuclei, but also other structures known as organelles. The two most important of these are mitochondria, used to burn oxygen to power the cell, and chloroplasts, used for photosynthesis. It is thought that these first existed as independent bacteria, but became incorporated into the eukaryotes in a symbiotic relationship.

Archaea territory – hot pools and travertine formation at the Mammoth Hot Springs in the **Yellowstone National Park**

The mitochondria even have their own DNA, much studied by geneticists as it is inherited in both sexes from the mother only. It is not recombined from male and female sources at each generation like the main, hereditary nucleic DNA, and so it provides an ideal mechanism for the study of heredity, all the way back to African Eve.

Eukaryotes may be more advanced than prokaryotes, but they are still only single-celled creatures. It was the development of multi-celled organisms which led to higher forms of life, where groups of cells could specialize to form skin, muscles and so forth. No trace of such creatures is present in the geological record for 900 or more million years after the proposed appearance of the eukaryotes.

Eukaryotes must have evolved from combinations of prokaryotes, but this is thought to have been a one-off development, because all eukaryotes – including all plant and animal forms – share the same basic structure. Some scientists think the event so rare that it may never have occurred at all on other planets in the universe which possess prokaryote life forms. It may mean that man is the only creature which can send radio beams into space.

****

Anyone who wishes to learn about geology is up against one, immovable fact – he is going to have to commit to memory the main divisions of geological time, such as the Carboniferous and the Jurassic, and the order in which these periods juxtapose. However, there is one very simple place to start, the great division between the Precambrian, which ran from the earliest times until 542 million years ago, and the Cambrian and subsequent eras which followed. This latter is known as the Phanerozoic, the "eon" of visible and abundant life. Very large parts of the surface of the earth – in North America, Australia, Scandinavia, southern Africa, northern Brazil and so on – consist of Precambrian shield or basement rocks, nearly all of them changed beyond recognition from their original form. These are like a book with only the chapter heading written in, because the first creatures with a hard calcite shell or skeleton – which could be used to decipher the words – did not appear until the Cambrian era. Fossils are like clocks in the rocks, and can be used to calibrate sequences of rocks around the world. The Precambrian does not have them, though there are other forms of radioactive clock which can be used to give an absolute date to these rocks. These are used to divide it into two, the Archaean, before

2.5 billion years, and the Proterozoic, 2.5 billion to 542 million years ago. The Archaean consists almost wholly of metamorphic rocks, but there are some sedimentary rocks within the Proterozoic.

An additional problem with the Precambrian is that it is too old to contain coal, oil or gas, which derive from abundant life, so it is of no interest to coal or oil company geologists. It is however well-endowed with minerals – for example the gold, copper and diamond mines of southern Africa are based on Precambrian rocks. However, there are just so many universities with graduate programs for geologists today that the Precambrian is coming under increasing focus. After all, why go and dig out the same trilobite or ammonite that has already been turned over by generations of geologists?

It is thought that the early earth contained no continents, just a vast deep ocean penetrated here and there by island arcs and their associated volcanoes. Today, about 40% of all the earth's crust is made up of continental (as opposed to ocean-floor) rock, so where did it all come from? In fact it came from the volcanoes. When two oceanic plates collide, one of them plunges into the mantle, to a depth of about 145 km/90 miles, taking a lot of water with it. At these depths the heat drives the water out from the subducting plate and into the mantle rocks above it. Here it has an unexpected effect on the hot rocks of the mantle – it lowers their melting point. Lava with minerals containing a preponderance of the lighter elements (oxygen, silicon, sodium, aluminium) begins to form. As the lava works its way towards the surface, the heavier elements such as iron and magnesium crystallize out and stay underground. The lava then erupts as rhyolite or andesite, or crystallizes underground where it forms granite or diorite. In both cases these rocks are not like the mantle from which they came. They are continental rocks, depleted of the heavier elements. Brand new continental rocks are being minted in this way in island arcs such as the Marianas right now.

Gradually the clumps of continental rocks formed small continents (like New Zealand today), and these grew by gathering other island arcs which were swept into them by crustal movements. In this way it can be seen that the thousands of islands of Indonesia will one day become part of a continent as the plate on which New Guinea and Australia sit ploughs inexorably northwards into Asia.

The remains of these early continents can still be found embedded in the great shields. They are known as cratons, and their roots are thought to descend several hundred kilometres into the mantle. Some cratons are Archaean in age, dating to at least 2.5 billion years ago; all are

Precambrian. The roots of the cratons are thought to be the most likely places for the creation of diamonds, as the mantle itself at depths of 150 kilometres/90 miles is simply too hot for this process.

The most typical rock of the Precambrian shield is gneiss, which in its extreme, most badly squashed form is known as granulite. Large, irregular formations of gneiss represent the metamorphosed cores of ancient island arcs. Curving around them in great welts are ancient metamorphosed sediments, originally deriving from the erosion of the island arcs. Also associated with them are greenstone (metamorphosed basalt) belts, where the rocks appear green because of the abundance of the metamorphic mineral chlorite, as well as other green minerals, epidote and hornblende These belts are thought to represent the lines along which the ancient proto-continents collided, and are frequently rich in precious metals – gold, silver, copper and nickel. An example is the Red Lake Greenstone belt in Ontario, which is gold-bearing. Other common shield rocks are granite, granodiorite, diorite and schist.

Life certainly existed in abundance in Precambrian times, even if the fossil record is scant. Such fossils as there are consist mainly of stromatolites (from the Greek words meaning "mattress rock"). These are layered structures – resembling large, round cushions – formed in shallow water by the trapping and cementation of sedimentary grains by microorganisms. They form laminated structures, with a top layer of cyanobacteria and anaerobic bacteria beneath them. The cyanobacteria excrete a sticky slime which traps grains of sand.

Threads and rods thought to indicate the presence of early stromatolite life can be found in very early rocks – mainly cherts – which have somehow escaped metamorphosis. The oldest of them date back 3.5 billion years. Although their authenticity is in doubt, there is evidence from another quarter, geochemistry, that life was present on earth at that time in any case. A well-known example of an ancient fossil-bearing rock is the Fig Tree chert of southern Africa.

Stromatolites provide some of the most ancient records of life on Earth. They were much more abundant on the planet in Precambrian times, reaching their maximum population about 1.25 billion years ago. They subsequently declined in numbers, which by the start of the Cambrian – over 600 million years later – had fallen to 20% of their peak. It is thought that they came to serve as food for grazing creatures, implying of course that such creatures did exist by over 1 billion years ago. Molecular clocks, which use the rate of change in DNA between species to measure the time back to a common ancestor, do yield dates going back this far for complex organisms. Modern stromatolites still exist at Shark Bay in Western Australia and in Lower California. In all the vast eons of their existence, they hardly seem to have changed, offering little to the palaeontologists.

That the Precambrian was a very different place to the modern earth, especially in its atmospheric chemistry, is indicated by the geology. It is thought that the earth's atmosphere at first included little or no free oxygen, but that the amount of this gradually increased by the respiration of bacteria. Cyanobacteria live by photosynthesis, using sunlight to break down carbon dioxide from the atmosphere into carbon, which they use for energy, and oxygen, which they release back to the atmosphere. Over the eons, they must have produced vast quantities of oxygen, but at first it was immediately taken up by other materials, which readily combined with such a reactive element. This resulted in the formation of (carbonate) limestones and ironstones, but after a certain point, oxygen was produced faster than it could be sequestered, and free oxygen began to appear in the atmosphere in large quantities. This major environmental change happened around 2.4 billion years ago.

The rising oxygen levels may have wiped out a huge portion of the Earth's anaerobic inhabitants at the time. Cyanobacteria, by producing oxygen, were essentially responsible for what was possibly the largest extinction event in Earth's history.

A distinctive feature of the Precambrian are banded ironstone formations (BIFs). A typical BIF consists of repeated, thin layers of iron oxides, either magnetite (Fe_3O_4) or less commonly haematite (Fe_2O_3), alternating with bands of chert. Banded iron layers were a common feature in sediments for part of the Earth's early history but became much more rare as atmospheric chemistry changed.

The formations are abundant from 2.6 billion years ago, and become less common after 1.8 billion years ago. The total amount of oxygen locked up in the banded iron beds is estimated to be perhaps twenty times the volume of oxygen present in the modern atmosphere. Banded iron beds are an important commercial source of iron ore, such as those found in the Pilbara region of Western Australia, and the Animikie Group in Minnesota.

It is thought that the banded iron layers were formed in sea water as the result of free oxygen combining with dissolved iron in the oceans, precipitating out as iron oxides. Conditions may have favoured sudden blooms of blue-green algae (as occur today) which came and went, causing the banding with chert, which is the distilled essence of millions of tiny creatures. It is assumed that the earth started out with vast amounts of iron dissolved in its oceans. Eventually, as photosynthetic organisms generated oxygen, the available iron in the oceans was deposited as iron oxides.

It is well-known that the continents of the earth were once all joined together into one super-continent known as Pangaea, in Permian times (299-251 million years ago). However, the paleomagnetic people have been busy, and believe that this happened several times before that. The supercontinent before Pangaea is known as Rodinia, and existed about a billion years ago. Latest opinion also favours a earlier agglomeration about 1.5 billion years ago. An igneous rock which dates from this era is the Scandinavian rapakivi granite, which has the appearance of sliced eggs floating in a dark matrix. One place to see this is in a bar at Paddington station in London, where it forms the counter.

The evidence for Rodinia includes igneous rocks which were deformed (metamorphosed) about the same time, 1.3-1.1 billion years ago, the Grenville Series. These rocks can be traced across Western Australia, Canada, central India, coastal Antarctica, and North America.

****

It is common knowledge that the earth has recently undergone a major glaciation – indeed we are thought to be only in an interglacial period now. One factor involved is the disposition of the continents on the surface of the earth. The current situation were a major continent – Antarctica – sits right over one of the poles greatly assists the formation of an ice cap. There is speculation that the earth has been subject to earlier glaciations when a similar combination of circumstances occurred. In fact ice ages seem to occur every two hundred million years or so. The most controversial of these is Snowball Earth, in which it is proposed that the Earth's surface became entirely or nearly entirely frozen in phases between 750 and 590 million years ago. If nothing else, the idea certainly has a ringing scientific name, in the manner of the Big Bang, the Selfish Gene and the Black Hole.

Interest in – and consternation about – Snowball Earth increased dramatically in 1998 after Paul F. Hoffman, professor of geology at Harvard University, found fresh evidence in the rocks of Namibia. Here he found deposits laid down on a carbonate platform in a warm sea. There is nothing unusual about this – limestones like this are being deposited on carbonate platforms in the West Indies today. Above them however was a layer of glacial till, and above that, a resumption of the limestones. Exactly the same sequence had already been noted by others in the Spitzbergen Archipelago in the Arctic. Here a late Precambrian carbonate sequence clearly laid down in a warm sea was found to be interrupted by a sequence of glacial till apparently dropped from floating ice.

Hoffman supported his hypothesis by incorporating data from carbon isotopes. As photosynthetic life prefers the lighter carbon-12 isotope, the sequestered carbon in rocks generally contains a relatively high proportion of the heavier carbon-13. However Hoffman found a relatively high carbon-12 isotopic signature in his "cap" carbonates, consistent with an absence of life.

He also suggested that at the melting of Snowball Earth, water would dissolve the abundant CO_2 from the atmosphere to form carbonic acid, which would fall as acid rain. This weathered exposed glacial debris, releasing large amounts of calcium. When washed into the ocean this formed limestones unrelated to any biological activity. He called these "cap carbonates", sediments found on top of the glacial till that gave rise to his Snowball Earth hypothesis. An alternative mechanism is the rapid, widespread release of methane, unrelated to the Snowball Earth. (Note, however, that the sudden release of large quantities of methane into the atmosphere has been proposed to explain any number of otherwise mystifying geological phenomena!)

The main objection to the idea is that it is so non-uniformitarian. Hoffman proposes a whole set of processes which cannot be observed today, such as the very difficult task of getting the earth to freeze at the equator. Also, some of "till" deposits are hundreds of feet thick and so hardly seem likely to have accumulated by falling off ice floes. The rates of deposition are non-uniformitarian. Turbidity currents and underwater landslides, which are relatively common, have been proposed as an alternative explanation. Something of a compromise is the idea of a "slushball" earth with a thin equatorial band of open water.

****

The most fundamental geological division in Britain can be traced between the mouths of the Tees and the Exe – known, appropriately enough, as the Tees-Exe line. To the north and west of this line, the rocks are generally old and hard – Palaeozoic or even Precambrian. To the south and east of it lie the younger, softer rocks of the later periods, clays, limestones, chalks and sandstones. That this geological division is also reflected in the history of the country goes without saying. Until the eighteenth century the rich plain of the south-east was the heartland of the country. Then for a time, during the Industrial Revolution and after it, the geological resources of the north and west – coal, iron, tin, limestone, china clay and abundant soft water amongst them – gave a period of economic ascendancy to that region.

For such a small area, the British Isles contain a tremendous variety of rocks, both in terms of age and type, but there are reasons for this. Firstly the islands lie at the edge of a continent, and so have been subjected to repeated and long inundations by the sea, which have left many important sedimentary rocks behind. Second, and related to this, the country has consistently found itself near the boundaries of tectonic plates throughout geological time, and in terms of geology, these are easily the most exciting places to be, what with all the subduction, vulcanism and mountain building. The last time Britain was in such a position was in the Tertiary, when an important volcanic province was created in Scotland, and even now, the nearest plate boundary is only a few hundred miles away, in Iceland.

In addition, it is thought that the southern part of Britain has drifted northwards from a position near the Antarctic Circle 500 million years ago. Hence there is evidence of rocks from both ancient glaciations and tropical forests (Coal Measure), Saharan-style deserts (New and Old Red Sandstone) and coral reefs (Carboniferous Limestone) as well as run-of-the-mill clays and sandstones.

The oldest rocks in the British Isles are metamorphic, forming the Lewisian Gneiss. This is widespread in the far north-west of Scotland and in the Hebrides, notably of course on the island of Lewis. The oldest formations have been found to be 2.9 billion years old, amongst the oldest rocks on the entire earth. (The oldest rocks of all are found at Isua in Greenland and are approximately 3.8 billion years old.) The Lewisian Gneiss has been studied in great detail and is now known to represent rocks formed over a period of 1500 million years, from 2.9 to 1.4 billion years ago – in fact a greater time than all the time which has followed. The rocks are immensely hard, shiny and crystalline. The Lewisian gneiss is an extension of the Laurentian (North American and Greenland) Shield, of which it was a part until about 200 million years ago.

The attention of the world was drawn to the Lewisian Gneiss by the geologist who discovered its properties one summer's day in 1883, one J.J.H. Teall, at a place called Scourie. He thought he had found a dolerite dyke within what was obviously an ancient gneiss. Afterwards he realized that the dolerite had itself been subjected to metamorphosis.

He then published a paper called "On the Metamorphosis of Dolerite into Hornblende Schist", which says it all, really. Ever since then the heavy boots of structural geologists, geochronologists, geochemists and paleomagneticists have beaten a trail to the famous Scourie dykes. Places like the far north-west of Scotland or Siccar Point in the Southern Uplands of Scotland may as well not exist for the general public, but for geologists they are the centre of the universe.

Another very old formation forms caps on top of the Lewisian, the Torridonian sandstone. This is dated to between 1000 and 800 million years old. This is the first example in Britain of an unconformity, a discontinuity where one rock lies on top of a completely different (and in this case very much older) sequence. Amazingly, much of the Torridonian is bedded horizontally, that is, it has not been folded in almost a billion years.

South-east of the Lewisian and Torridonian province is a band of Cambrian rock, and south-east of this again lie the metamorphic rocks of the Moine formation. The original deposits may date back to the time of the Torridonian Sandstones, between 1200 and 870 million years ago, but these rocks have since been subjected to three separate periods of metamorphism, the last around 430 million years ago. The relationship between the Cambrian formations and the Moine Schists will be discussed in Chapter 7.

South-eastwards again across the Great Glen fault (which contains Loch Ness) lies the Dalradian formation. These rocks were buried deep in the Earth's crust where they were then deformed and metamorphosed through exposure to high temperatures and pressures, to form rocks such as schist and gneiss. All fossils were destroyed in this process. This occurred during the Grampian orogeny (period of mountain building), about 500 million years ago. Since then, through uplift and erosion, they have been brought back up to the surface where they are now the most commonly found rocks across the central Highlands. These rocks include the Grampian series and the majority of them predate the Cambrian, some being as old as 750 million years. The geological map of this area is spattered with large red blobs, and on geological maps, those mean granite intrusions. Towards the top of the Dalradian and dated to about 600 million years ago is the Port Askaig Tillite, ancient boulder clay left behind by retreating glaciers. There is evidence here of over forty separate ice advances. The tillite is thickest on the Isle of Islay at 750 m/2400 feet. The interesting thing is that this part of Scotland is thought to have been located within ten degrees of the equator in this period, so here is evidence from Britain of the famous Snowball Earth.

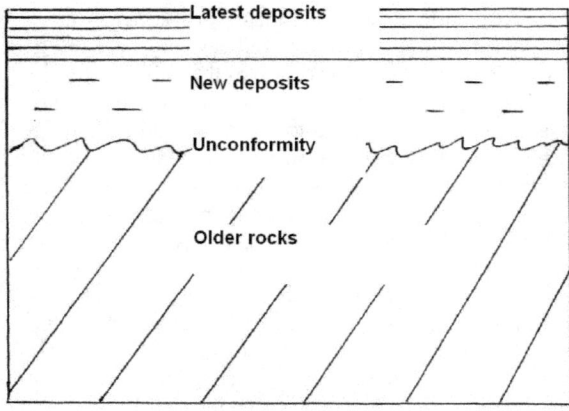

An unconformity, as is found between the newer Torridonian Sandstone lying on top of the Lewisian Gneiss. A famous example is found at Siccar Point in Scotland, where vertical beds of Silurian slates and grits are overlain by horizontal beds of Devonian sandstones, which are about 50 million years younger.

****

It is a fact that the vast bulk of the Precambrian is a palaeontological desert, but in the post-war era, something new was found to cast a dim light on these difficult rocks. Fifty million years after Snowball Earth, if it existed, we come across the first real fossils. The Ediacaran fauna of soft-bodied creatures was first discovered in Australia in 1946 by a man called Reg Sprigg (hence "*Spriggina*") in strata immediately predating the Cambrian. These fossils are best seen in low sunlight at dawn or dusk and Sprigg had difficulty in persuading the geological world that they were anything more than fortuitous inorganic markings. He eventually gave up in disgust and resigned his post at Adelaide University. He went on to found his own surveying, then oil company, and made his fortune in the gathering excitement about the fossils he had found.

The shadowy Ediacaran fauna, containing jellyfish-like animals resembling quite large floating mats, has since been identified elsewhere, but modern research now divides it into two. The type sequence – to which all other sites are compared – for the earlier ecology is at Mistaken Point in Newfoundland. In this place, a deep sea bottom, deposition continued straight through into the Cambrian period, where the Precambrian fauna has entirely disappeared. The typical animal is a very simple one, *Charnia*, which looks like a fern and which could not move. Within England, Precambrian rocks of this type are found at Charnwood Forest in Leicestershire. These are much younger than the Torridonian Sandstones, and indeed contain examples of *Charnia*. Charnwood Forest is clearly visible from the M1.

The Ediacaran fauna of Australia lived in shallow water. It contains much more sophisticated creatures than Mistaken Point, including *Spriggina* itself. This shows bilateral symmetry, segmentation, a head and a tail, a digestive system, the ability to move, and probably a mouth; in fact somewhat resembling a trilobite without a shell. Another species, *Dickinsonia*, the floating oval bathmat, was more than one metre (three feet) long and also exhibits bilateral symmetry. Incidentally, the existence of jellyfish fossils – which look remarkably similar to modern species, most of which are still very primitive creatures – indicates something unusual in preservation conditions in the late Precambrian, as they are not found in later rocks, though obviously the creatures are still with us.

Another outcrop of the Precambrian is the inlier of Longmynd, a well-known hill west of Church Stretton in Shropshire. (An inlier is an older outcrop surrounded by younger rocks. An outlier is a younger outcrop – normally a hill – surrounded by older rocks, away from the main outcrop of the younger rock. In this way, the free-standing Pendle Hill in Lancashire, home of the witches, is an outlier of the main Pennine range.) The Malvern Hills to the south are also Precambrian, and there are rather older Precambrian rocks in Anglesey and on the Lleyn Peninsula opposite, and in the Channel Islands. The Welsh rocks form the Monian Group (from *Mona*, the Roman name for Anglesey, and not to be confused with the Scottish Moinian Group). This contains a remarkable sequence called the Gwna Mélange, a chaotic mixture of boulders – limestone, pillow lavas, quartzite and even jasper – embedded in a matrix of mudstone and sandstone. It is thought to have been formed by a powerful submarine slide, and is an example of an olistostrome.

In fact Precambrian rocks underlie England, at depth; but England is situated at the edge of a continent, and so has been subject to repeated and long periods of inundation by the sea as the earth's crust and sea levels have periodically fallen and risen. In the interior of the continental masses, far away from the sea, Precambrian rocks are found at the surface to a much greater extent.

However the British Isles is far from the best place in the world to study Precambrian rocks. Those of North America are not only very widespread but have been extensively studied, as they outcrop widely at the surface, especially in Canada. Here the continent is divided into nine Precambrian provinces, each of which has been welded onto a preceding block in crustal movements of long ago. The oldest province is the Slave, in the far north-west of Canada, dated by zircon crystals to ages between 3.9 and 2.5 billion years old. Another is the Superior, 3.2-2.6 billion years old. The whole heartland of the continent was formed in this way. Some of the greatest unconformities in the world are found here. The rocks either side of the Cheyenne belt in Wyoming are thought to be a billion years different in age, and the younger ones are still 1.8 billion years old. By contrast, the coastal states of the east, west and south lie upon foundations built up in the Phanerozoic, that is, after the start of the Cambrian.

A number of geophysical tools are available which have proved of great value to Precambrian specialists, as the rocks themselves are largely buried under later sediments. These include seismographs, gravimeters (which measure gravity) and magnetometers (which measure magnetic fields). In North America, the most useful has been the magnetic survey, which can be conducted from the air. The effect of it is to strip away the Phanerozoic cover to give a good idea of the underlying geology. When controlled against field samples taken from drilling it can now produce an approximate map of the Precambrian basement. (Some field workers, notably Australian, regard these maps as "fantasy geology".)

One feature which shows up strongly on this map is a mysterious string of granite batholiths running across 4,000 km/2,500 miles of country, all of about the same age – 1.45 billion years, and so known as the "1450 plutons". They are mysterious because nowhere else in the world is such a large number of granite batholiths known, and because they do not appear to be associated with any orogeny – granite batholiths elsewhere in the world ARE a characteristic feature of mountain building.

The composite magnetic anomaly map of North America also features the Midcontinental Rift, where iron-rich basaltic lavas give off a strong magnetic signal. This stretches from Iowa to Lake Superior, where it forms a three-way junction with two other rifts – a structure not dissimilar to the three-way modern junction at the oceanic end of the Red Sea, where rifts from East Africa, the Red Sea and the Gulf of Aden (offshore Saudi Arabia) meet. The difference is that this rift is dated to 1.108 – 1.086 billion years ago. Amidst vertical movements of a thousand metres (3,000 feet) or more, proto-North America threatened to split apart. However, the rift stabilised and filled up with the outpourings of basalt which now show up in magnetic surveys.

****

The Church Stretton area is very complex geologically, at least by English standards. The town itself lies south of Shrewsbury. To the west of it is the Longmynd, to the east the Stretton Hills, and to the south the Wrekin, all late Precambrian. In the valley below Longmynd and passing close to the town is a major fault, the Church Stretton fault, passing right through Shropshire from south-western Wales in a SW-NE direction, and thought to have been active throughout Phanerozoic times – but it is far from being the only fault. Another nearby one (the Pontesford-Linley running parallel to the north, and equally ancient) registered an earthquake of a magnitude of 5.1 on the Richter scale in April, 1990. During the Silurian period around 420 million years ago the limestones of Wenlock Edge, just to the east of Church Stretton, were formed as tropical coral reefs. Some of these limestones can be found in the Church Stretton valley, having been pushed down into it along the Church Stretton fault. In fact the whole area has been horribly jumbled up by the faults.

When I started as a student at Oxford, I was required to take an exam – a "Prelim" – after just two terms; anyone failing the exam had to retake the part they failed, and keep retaking it until they passed, or leave the university. This exam included a paper on geological maps. In my first two terms we had a class dealing with this subject every Saturday morning, and they soon began to strike fear into my fellow students. Essential to geological maps are the ideas of "dip and strike". The dip is easy – the angle at which an outcrop of rock dives into the ground. Strike is really just a compass bearing giving the orientation of the rocks as they enter the ground. If the bedding planes of the rocks are lying straight east-west, for example, the strike would be N90E –

ninety degrees east. Dip and strike can also be applied to faults, and geological maps incorporate these and other more arcane symbols for faults. It wasn't easy to get out of bed as an 18-year old student, especially on a Saturday. I went to one session and a fellow from Keble College called Hartshorn seemed really on top of the job, curse him! Nobody else seemed to have much of a clue. So despite dire warnings, I stopped going to the mapping sessions. Come the day of the exam, it would be all right. I certainly learnt a lot of other geology, but was never keen on practical things like these mapping sessions. Anyway the class maps were always of easy areas – mostly the local area around Oxford, as I recall.

You can imagine my horror when I turned up for the exam itself and saw the map in front of me – Church Stretton. I was well aware that this was a most difficult area. The examiner from the Geology department must have had a wicked sense of humour. Students with two terms of Geology behind them were expected to draw a section of the rocks as they passed underground, from the Longmynd, across the faults and heading west. What he made of my effort, I really don't know. What I do know is that Hartshorn got a distinction in the Prelims, one of only two awarded amongst seventy students.

****

Before looking at the Phanerozoic rocks of Britain, a note is required about the nomenclature. In recent years some rocks have been reclassified or renamed, and the traditional names, known to generations, no longer appear on the maps of the British Geological Survey. These names do however continue to appear in books such as this one, written for a general readership rather than for specialists, and they are likely to remain in use as they represent the easiest way of classifying these rocks. For example the Permian Magnesian Limestone is now known as the Cadeby and Brotherton formations, with numerous other small formations defined around it. Similarly the term "Tertiary" for the era between the Mesozoic and the Quaternary is no longer found on maps of the British Geological Survey. Another term which has been replaced is "Hercynian", used for the mountain-building era at the end of the Carboniferous, and now called the Variscan.

In common with other writers, including Professor Richard Fortey, whose books I heartily recommend, I intend to continue with the traditional nomenclature.

Chapter 7 – Lower Palaeozoic

Phanerozoic time is divided into four great eras, of which the first, the Palaeozoic, is by far the longest, lasting from 542 to 250 million years ago. In turn it is divided into six periods, each still of great length, and within these periods are many smaller subdivisions of time of interest to professional geologists. The term *Palaeozoic* means *ancient life*, compared to *Mesozoic, middle life*, and *Cenozoic, new life*.

The Cambrian is the first period of the Palaeozoic era, lasting from 542 to 489 million years ago. This was named after Cambria, the Roman name for Wales, by Adam Sedgwick (1785-1873). Sedgwick himself had trained as a mathematician at Cambridge. He grew up on a farm near Dent in North Yorkshire, where a large stone fountain commemorates his name to this day. On his appointment as Professor of Geology at Cambridge in 1818, he noted "Hitherto I have never turned a stone; henceforth I will leave no stone unturned"!

The Cambrian Period marked a great change in life on earth as sea creatures with hard shells, which fossilized readily, became common. This diversification of life forms was relatively rapid, and is termed the Cambrian explosion. All modern groups of species, known as phyla, date from this period. Altogether, thirty-seven body plans emerged, one or other of which has been inherited by every animal on the planet today. A phylum (plural phyla) is a major group of animals. One such is the arthropods, invertebrate animals having an exoskeleton (external skeleton), a segmented body and jointed appendages. All insects are arthropods, as are the arachnids (spiders) and crustaceans. The crustaceans form a subphylum within the arthropods. These are mostly free-swimming sea creatures such as shrimps, krill and crabs, but also barnacles and land-based animals such as woodlice. Crustaceans are first recognised from the Cambrian. (Note: not to be confused with cetaceans, which are marine mammals – whales and dolphins.)

The most famous and characteristic arthropod fossil of all is the trilobite, first appearing in the Cambrian. Trilobites flourished throughout the lower Palaeozoic era before beginning a drawn-out decline when, during the Devonian, all trilobite orders but one became extinct. Trilobites finally disappeared in the mass extinction at the end of the Permian about 250 million years ago. Hence the trilobites were among the most successful of all early animals, inhabiting the oceans for over 270 million years. 17,000 species are known! There are so many, you could spend a lifetime studying nothing else, and indeed, some people do exactly that! All of them feature the three distinctive lobes – cephalon (head), thorax (body) and pygidium (tail). They were also the very first creatures which were known to have developed eyes, and very unusual, sophisticated eyes at that. These are composite eyes of many lenses, not made from soft tissue but from calcite crystals.

A small **trilobite**

Another Cambrian fossil group is the brachiopods, still found in the seas today. The name means "Arm foot" because in living species a fleshy arm can be observed which anchors the creature to the sea floor.

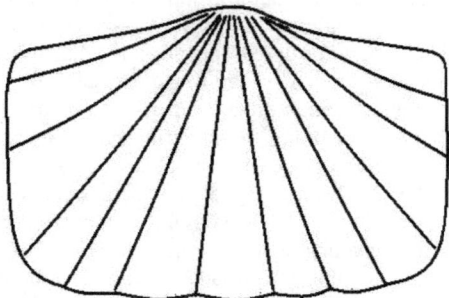

Brachiopod

Brachiopods are bivalves, having two shells which close together like a an oyster, and the bivalves in turn are part of one of the great groups of invertebrates which emerged at the start of the Cambrian, the molluscs. The other mollusc groups are the gastropods ("stomach foot"), which have twisted shells and which include garden snails and whelks today, and the cephalopods, which include the Cambrian nautiloids, and the later squids and ammonites. These have partitioned chambers inside the shells, the squid-like animals actually living in the last and largest partition.

The Cambrian has a very famous *lagerstätte*, a site where the soft parts of organisms are preserved as well as their more resistant shells. This is the Burgess Shale, located in the Canadian Rockies of British Columbia. Its existence means that our understanding of Cambrian biology is greater than that of some later periods. At 505 million years old it dates from the Middle Cambrian. The rock unit is a black shale.

The Burgess Shale was discovered by the American palaeontologist Charles Walcott in 1909. Many of the animals he found appeared to have bizarre anatomical features and only a slight resemblance to previously known animals. Examples include *Anomalocaris* ("abnormal shrimp") and *Hallucigenia*, which was originally reconstructed upside down. It was many years before *Anomalocaris* was eventually pieced together as one animal – it had been classified as three separate species – and made to look a little, if only a little, less anomalous. In fact it was a large predator, up to nearly two metres (six feet) long, and the different parts of it each looked like separate animals. The Burgess Shale was the subject of a work of popular science, *Wonderful Life* (1989) by Stephen Jay Gould.

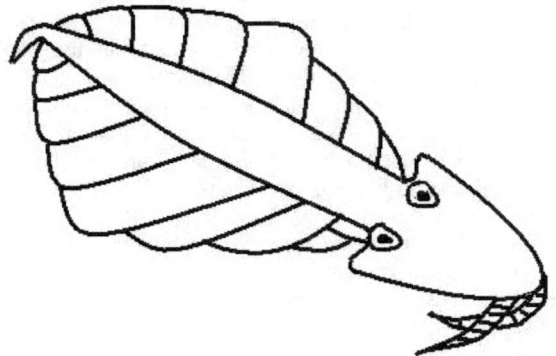

"Carrie" the Anomalocaris

The discovery of the Burgess Shale meant that the Cambrian oceans must have been inhabited by a hitherto little-known fauna, which must also have been present elsewhere across the world. Indeed, *Anamalocaris* has since turned up in China, as has the first known vertebrate, named *Myllokunminga*, a very inconspicuous marine creature thought to have had a backbone made of cartilage. The best Cambrian site in China is found in Yunnan in the south, at a place called Chengjiang. Here are more fossil beds in a fantastic state of preservation. They date from the lower Cambrian, 525-520 million years ago. They are comparable to the Burgess Shale, and though about ten million years older, contains the same startlingly original animal groups. On both sites, arthropods dominate. It was once thought that the trilobite was the only arthropod dating from this era, but it is now clear that there were many others.

However, the fact is that Cambrian fossils – especially trilobites, such as the large marker *Paradoxides* of the Middle Cambrian – are rare. It is possible to spend days at a time in the Cambrian rocks of North Wales or Shropshire without finding a decent fossil, whereas in say the Middle Ordovician rocks of Shropshire, they turn up in minutes, if not seconds.

****

As the name betokens, Wales has important Cambrian formations, laid down as sediments in the peripheral seas of a deep ocean known as

Iapetus. (This is the name of the Greek god who was the father of Atlas, after whom the Atlantic takes its name – the Iapetus Ocean used to be called the Proto-Atlantic.) These muds and silts were later altered by intense pressures, and turned into the slates which have been quarried for centuries in north Wales, notably around Llanberis and Bethesda. Cambrian slate became the roofing material of choice throughout Britain from the middle of the nineteenth century. It is often possible to use it to date property from this era, slate tiles replacing stone and other materials after the arrival of the railway in any given locality. The metamorphism of the shales into slates in most cases destroyed the fossils in the rocks, because the slate cleaves along the line of metamorphic pressure, not along the original bedding planes, which would normally be the best place to find fossils. Only in the unusual locations where the line of cleavage and the original bedding planes coincide can the odd fossil be found.

The Llanberis slate belt occupies a relatively small area, north of the Snowdon massif. South-west of Snowdon in the "armpit" of North Wales is another Cambrian formation, the Harlech Dome. There are more Cambrian rocks further south in Wales. Between Barmouth and Dolgellau, basalt sills are found, shot through with gold-bearing quartz veins. The wedding rings of the English royal family are famously made from this gold.

During the Cambrian, Scotland lay south of the equator (and separated from England by thousands of miles of ocean). Parts of it formed the floor of another sea of the Iapetus ocean, just off the southern edge of Laurentia. (Named after the St Lawrence River, "Laurentia" is the name given to the early continent of North America and Greenland, which in some shape or form has persisted through to the present day.) The Cambrian in Scotland is located in the north-west, overlying the Lewisian Gneiss on the landward side. The rocks outcrop mainly around Assynt, south of Durness and on the Isle of Skye. This sedimentary succession shows a progression from tidal sands to limestones deposited on the marine shelf (the Durness Limestone, the first British rock to be composed mainly of the remains of sea shells, and noted for its limey fertility). Fossils of trilobites can be found in these rocks. The most famous rock is the pipe rock, which shows abundant vertical fossil burrows. Another sequence is a sedimentary quartzite – a sandstone indurated to quartzite through its great age. Like other quartzites, this rock is so tough that it does not weather to produce soils, but only a mass of straight-edged blocks which form some of the most difficult walking country in the whole of Britain.

****

The Ordovician is the next geological period, running for 44 million years from 488 million years ago. It was separated off from the Cambrian, which precedes it, and the Silurian, which follows it, after one of the longest-running of four great Victorian geological disputes, two of which involved the same man – Sir Roderick Murchison (1792-1881). He had served as an officer in the Peninsular War before marrying and giving up the military, eventually in favour of the new and expanding science of geology. Knighted in 1846, Murchison was for many years the President of the Royal Geographical Society, which was an enormously influential – and distinctly aristocratic – body in the nineteenth century.

In 1835 Murchison and the Cambridge professor Adam Sedgwick presented a joint paper, under the title *On the Silurian and Cambrian Systems, Exhibiting the Order in which the Older Sedimentary Strata Succeed each other in England and Wales.* This was the foundation of the modern Palaeozoic timescale (the more recent geological periods had already been named by Lyell). When traced away from its source area in south Wales, Murchison's Silurian series came to overlap Sedgwick's Cambrian sequence, provoking furious disagreements. Sedgwick, who had humble start in life, found himself outwitted by the aristocratic Murchison. He watched appalled as Murchison's Silurian colours crept up the map of Wales, reducing his Cambrian sequence to a rump. The dispute also spread to English formations. Sedgwick later wrote to Murchison:

"Your nomenclature of the older English rocks is false for the simple reason that below your true Silurian groups (beginning with the Wenlock Shale) your original and typical section and your order of superposition are false. My order of superposition was not false in any essential part..... You have acted contemptuously, unjustly and falsely towards me. I cannot smooth over the matter by the shallow gloss of vulgar courtesy or by abuse of the name of friendship."

It was left to the next generation of geologists to resolve this dispute. Here we meet Charles Lapworth (1842-1920), who started his career as an enthusiastic amateur – a schoolmaster – and ended it as Professor of Geology at Mason College (which became Birmingham University). He proposed that the disputed strata should be placed in a period of their own, based on their distinctive fauna of graptolites. These fossils resemble floating razor shells or tuning forks. Their name means "stone writing", and their white or grey remains can be found on the bedding planes of ancient shales. Lapworth called the new period the Ordovician, after the Welsh tribe, the Ordovices (who gave such trouble to the Romans that they were slaughtered *en masse*.) His new period was eventually accepted worldwide.

Although trilobites and brachiopods are much better known, eventually graptolites were to be used to establish the correct stratification of the Lower Palaeozoic. This is because they are a good zonal fossil, a zone being a unit of stratigraphical time. The problem is to correlate the same zone in different types of rock, where there are different fossil assemblages, or no fossils. A whole assemblage can be used in zoning, or a single fossil type with a recognisable evolutionary pattern. Brachiopods for example favoured continental shelves where limestones were forming. Graptolites have a wider distribution as they were a floating, planktonic form which favoured deeper waters in which black shale was formed.

Arriving in Southern Uplands of Scotland, Lapworth at first found little but unprepossessing and unfossiliferous greywackes, hard sandstones of poorly-sorted angular fragments set in a finer matrix. Then he came upon something much more interesting, bands of interbedded black shales. It was these rocks that contained the graptolites. Recognizing that great truth of palaeontology, that you have to make the best of what you have got, at first he could make no sense of the graptolites. They failed to show the expected evolution between separated outcrops.

Lapworth was eventually reduced to mapping a whole region. This caused him to realise that repeated beds of black shale a maximum of only 150 metres (500 feet) thick amongst beds of greywackes 8,000 metres (26,000 feet) thick (!) were in fact the same bed, ruptured and folded – yet not metamorphosed – containing the same graptolites. The greywackes were nowhere near 8,000 metres thick, but folded in the same way with repetitions of sequences. At a site called Dob's Lin, near Moffat in the heart of the Southern Uplands, Lapworth finally unlocked the arcane history of the rocks. He had succeeded in the definitive mapping of the boundary between the Ordovician and the Silurian. As graptolites floated freely on the oceans, this boundary is applicable across the globe, and his work is of international importance – Dob's Lin is now a world-famous site for geologists. The site is difficult to interpret because the younger Silurian deposits to the south appear to lie beneath the older Ordovician greywackes to the north. Lapworth attributed this to intense folding. The modern interpretation of the structure is that the rocks are indeed folded, but also divided by thrust faults. The strata were deformed in this way by a subducting oceanic plate pushing beneath them, pulling the Silurian rocks downwards and bending them over. The thrust faults arose from the compression of the crust. This type of structure is known as an accretionary wedge. Accumulated oceanic sediments are pushed onto an area of continental crust in a configuration which is triangular, or wedge-shaped, in vertical profile, with the narrow end of the triangle disappearing into the depths. There are several more modern examples round the world, for example on the coast of Chile.

The lower boundary of the Ordovician, given here as 488 million years ago, has been subject to some revision. Some authorities (see Peter Toghill) now place it as early as 510 million years ago, which makes the Ordovician 66 million years long (71 million according to Toghill, 510-439). Certainly the last epoch of the Cambrian, the Tremadoc (510-493 million years ago on Toghill's dating), is now placed in the Ordovician because the trilobite and other marine fossils within it are similar to later Ordovician forms. Sedgwick would not be pleased!

The world map of the continents looked very different in the Ordovician. What we now know as the southern continents formed a single continent called Gondwana. This included modern Africa, South America, Australia, Antarctica, Arabia, India and Madagascar. The mini-continent Avalonia (named after a site in Newfoundland) separated from Gondwana and began to move towards the two northern continents, which are known as Baltica and Laurentia. Newfoundland, Wales and Spain share a common early Ordovician trilobite fauna containing *Neseuretus* and *Ogyginus*. These must have flourished in the seas around Avalonia. The Precambrian basement rocks underlying the England are called Avalonian on British Geological Survey maps, but "England" was probably above sea-level in this period. Avalonia was a microcontinent, in the manner of New Zealand today – a small piece of continental crust with an independent history.

Precambrian basement quartzite and gneiss formations at the mighty cliffs of Slieve League, Donegal, Ireland

One of the most mysterious Ordovician fossils, at least for many years, was the conodont. First appearing in the late Cambrian, these became very common fossil finds, because they remain behind when limestone is dissolved in acetic acid (they are made of calcium sulphate rather than calcium carbonate, which does dissolve). A few millimetres long, they appeared to be some form of teeth or jaw apparatus. Not until the 1980s, and after many fanciful reconstructions, were impressions of their soft parts found, complete with conodonts attached at one end. They are indeed teeth, and belong to something that looks like a sand eel – a marine vertebrate.

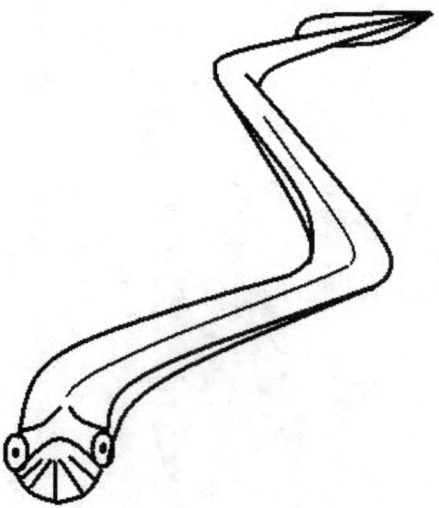

Connie the conodont

Conodonts have proved to be of great importance to geologists, because of their long history, from 512 to 208 million years ago, common occurrence and widespread distribution. One feature of these fossils which was observed by an American geologist called Anita Harris is that their colour changes according to the temperatures to which they have been subjected, from faint yellow to amber, darker brown, black, white and finally crystal clear, in a sequence of rising temperatures. This feature is of great interest to oil companies, because oil will only form where the rock temperature has reached 50 degrees Celsius, but no more than 150 degrees; beyond that, the oil is destroyed. At 120 degrees, the perfect temperature for oil preservation, the conodonts are a golden brown. So there is no point drilling for oil where the conodonts indicate temperatures above 150 degrees. Other mineral geologists, however, are only interested in the hotter places, where deposits of copper, iron, silver and gold might found, the sites of ancient hydrothermal springs.

Another creature which appears in the fossil record of the Ordovician is the horseshoe crab, dating from 475 million years ago. It has proved to be one of life's great survivors and still thrives to this day. It is an arthropod which actually has blue blood, coloured by copper. Another important group of fossils, the bryozoans, also made its first appearance in the Ordovician. These are tiny animals which form colonies of sea mats, bushes and fans.

At the end of the period there was a mass-extinction event, the first in the Palaeozoic. This was probably caused by the ice age that occurred at the end of the Ordovician, the first of only three in the Phanerozoic era. The ice sheets were centred on what is now North Africa, including Morocco, then part of Gondwana, and lying directly over the south pole. (Note that the fact that there was a mass extinction at this time does not seem to have led either Sedgwick or Murchison to draw a period boundary here.) All this brought the curtain down on the first great phase in the diversification of marine life – about 60% of all species were wiped out.

Within the British Isles, Snowdon and its vicinity form a syncline of Ordovician rocks which are partly volcanic. A syncline starts out in life as a downward fold in the surface (an anticline is an upward fold), but the processes of erosion and uplift can leave synclines lying higher than anticlines. Snowdon itself is largely formed of volcanic ash (tuff) and ignimbrites with some sedimentary rock and igneous intrusions. It is thought that the Snowdon area lay close to a zone of subduction where massive andesitic vulcanism was taking place. Huge amounts of ash were blown off and settled in the sea, entombing animals lying on the sea bed. These can now be found as fossils, right on the top of the mountain. The neighbouring Cader Idris is also largely formed of Ordovician igneous rocks. Roofing slates of the highest quality are quarried from metamorphosed tuffs of north Wales.

Ordovician rocks also outcrop in south-west Wales, where the noted tuning-fork graptolite *Didymograptus* is found. Extending eastward from the coast they form the Preseli Hills. It was from these hills that the famous bluestones of Stonehenge were quarried. This rock is a spotted dolerite. In England Ordovician rocks are found in Shropshire, most notably at the Stiperstones ridge, made of tough quartzites.

The most important area in Britain for Ordovician rocks is the Southern Uplands of Scotland, already described. There are more rocks of this age in south-west Scotland at Ballentrae on the coast of Ayrshire. Included in this series are ophiolites, sections of the oceanic crust which instead of being subducted have somehow ended up on the land.

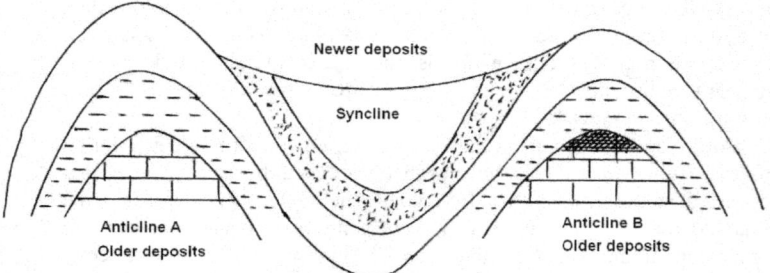

Anticlines and synclines. Here the anticlines A and B are in their original form, but the syncline has been partly eroded and filled with newer deposits. Anticlines are of great interest to oil and gas geologists because they trap oil and gas rising through permeable rock such as chalk when it is capped with a permeable layer such as clay, as in anticline B.

****

At the end of my first year at Oxford, I considered changing subjects, from Geography to Geology. One of the Geology lecturers suggested that I get used to things by going on their summer field trip – a week in Snowdonia. Well, it was a surreal experience. I don't think there's a great deal wrong with me, socially speaking. We all have our faults, I no less than others, but certainly, in this group of 20 or so people, I was a stranger. On the second day, we climbed Snowdon, and I had the distinction of finding a trilobite. On the third day, we went up Glydr Fawr, which is only a few feet lower than Snowdon, and which looks much the same – this seemed unnecessary. On the fourth day, I went home.

The geology lecturer tried to persuade me to stay, but I was adamant, because in those three days, not a single geology student had offered a conversation. Every attempt that I made myself to start a conversation was met with little more than a quick comment, a nod or a smile before the student hurried on – as if I had some dreadful contagious disease. Would this have happened in Italy?

****

The Silurian period is defined as extending from approximately 443 to 417 million years ago. As noted, this period was first identified by Roderick Murchison during the early 1830s, based on fossiliferous strata in south Wales. The Silures were a Welsh tribe of the Roman era. Murchison published the definitive account of the period in his *Silurian System* of 1839, one of the great early works of geological science. It is surely Sedgwick's revenge that the later designation of the Ordovician period has had exactly the same effect on Murchison's Silurian as the earlier Silurian had on the Cambrian, shrinking it to a fraction of its former coverage on the map.

The great ice caps of the Ordovician receded in the Silurian. In this period, recognizable jawed and bony fish first appeared, as did the first primitive land plants.

Marine deposition continued in northern and western Britain. This was similar to the Ordovician, but with markedly reduced volcanic activity and obviously a new set of fossils – new groups of graptolites (for deeper sediments), and brachiopods and trilobites (for shallower sediments). Evidence that the southern part of the country now lay in Caribbean-style warm, tropical waters (probably around latitude 25 degrees South) can be found from the coral reefs of the Much Wenlock limestone of Shropshire. The main evidence for volcanic activity can be found in repeated layers of bentonite, a rock thought to represent windblown ash from distant volcanoes.

Towards the end of the period, the part of Baltica containing southern Britain (Avalonia) had moved from a position near the Antarctic Circle at about 60 degrees south, 4500 km/2800 miles northwards over a period of a hundred million years. This represents an annual rate of 4.5 cm, equivalent to modern tectonic plate movements. It used to be thought that the northern continents of Baltica and Laurentia then collided, folding up the coastal sediments which had been accumulating between them ever since the Cambrian. More recent evidence, both on the ground and from seismographic surveys, indicates that the collision as it affected North American geology was between Laurentia and a subducting oceanic plate, and not Baltica.

In any event, this event is known as the Caledonian orogeny, a phase of mountain building that created ranges stretching from the eastern United States, through Newfoundland to Greenland, Wales, the English Lake District, Northern Ireland, Scotland and Norway. The chains of mountains produced must have rivalled the Andes in length – the section from Northern Ireland to the top of Norway alone spans ten degrees of latitude. The Caledonian mountains of Scotland, with their south-west to north-east trend, are really the northward extention of the Appalachian mountains of North America, and the southern extention of the coastal mountains of Norway, which both share the same trend.

"Orogeny" is a term which means mountain-building, but the mechanics of it were redefined after the discovery of plate tectonics. It used to be thought that the sediments accumulating off a land mass created a geosyncline, or sinking fold within the crust of the earth. A modern version is the North Sea, which is steadily building up sediments washed down by the great rivers of northern Europe – the Rhine, the Elbe, the Thames, the Seine and so on. After a long period the land on either side of the geosyncline was thought to move together to crush the accumulated sediments into mountain chains. It is now realised that the movement is actually caused by one tectonic plate crashing into another, and that only one plate may be moving. Hence the Alps are caused by the northward movement of the African plate into the European plate, which is itself stable. This means that the folds of rock or "nappes" (a term taken from the French word for a (crumpled) tablecloth) form a northward sequence, in some cases tumbling over and on top of one another.

Towards the end of the Silurian, as the mountains began to rise in the north and west, Old Red Sandstone – until quite recently generally regarded as a purely Devonian formation – began to be laid down. This is a terrestrial deposit. Meanwhile, the old-style marine sedimentation of greywackes and graptolitic shales continued to the east. A new epoch was eventually added on to the end of the Silurian period, named after a site in the Czech Republic, in recognition of this. The deserts which produced the Old Red Sandstone eventually replaced this as well, so the appearance of this formation is known as a diachronous event – the same geological event occurring at different times in different places.

Within Britain the Old Red Sandstone lies unconformably on the rocks below it. This is not nowadays thought of as the best place to define a period boundary (in this case Silurian/Devonian) as it inevitably represents a gap in time, possibly millions of years. Period boundaries (which have to be recognized internationally) should be defined in areas of continuous deposition, as was the case for Lapworth's Ordovician/Silurian boundary in the Southern Uplands.

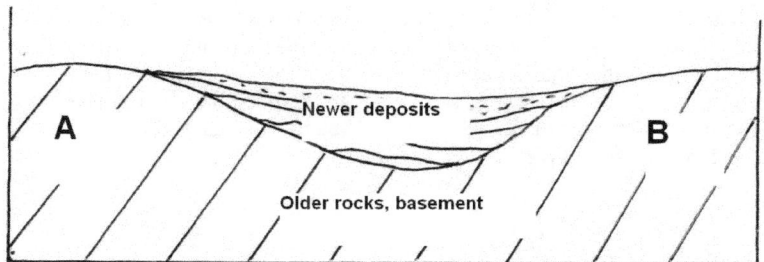

The old concept of a geosyncline was that newer deposits would be crushed into a mountain chain when continental blocks A and B moved together. It is now known that A and B represent separate tectonic plates and that only one of them is likely to be moving.

****

The Caledonian orogeny was the first to affect the British Isles in the Palaeozoic. It had a profound effect on the rocks of Scotland, Ireland and Wales, for large parts of those countries contain rocks created during the first three periods. The intensity of the Caledonian earth movements decreases southwards. The Lake District and North Wales do not show the same degree of heavy folding or metamorphosis that is found in Scotland. In fact the Grampians were used to define the degrees of metamorphism as it affects rocks of the same original composition, in a sequence from slate to phyllite, mica schist, garnet schist and gneiss, showing the classic metamorphosis from andalusite to sillimanite and kyanite.

****

The geological basis of New York city was laid down in the Caledonian orogeny, or the American version of it, the Taconic. There are three principal formations represented, the Manhattan Schist, the

Inwood Marble and the Fordham Gneiss. The first two were metamorphosed from relatively recent sediments at the time of the orogeny, 450 million years ago, but the Fordham Gneiss ("granite gneiss"), caught up in the same earth movements, is thought to be a billion years old. These three formation, interleaved together, form bedrock which is capable of supporting the gigantic skyscrapers which form two clusters, three miles apart, on Manhattan Island. They also made the excavation of the city subways such hard work as compared to London or Paris. However, there are also large amounts of recent glacial till and other glacial debris – outwash fans and moraines – which coat large areas of the surface, sometimes to considerable depth. The lower parts of the city such as Greenwich Village, SoHo and Chinatown lie on top of these deposits.

Chapter 8 – The Geology of Scotland and the Lake District

As the geology of Scotland and the Lake District is predominantly Lower Palaeozoic or earlier, it would be useful at this point to see how the pieces fit together in these two areas. Scotland in particular has been a place where the principles of geology were first established, some of which are illustrated here.

Scotland falls geologically into four distinct provinces, the Precambrian north-west, the Highlands, the Midland Valley and the Southern Uplands. The Highlands lie to the north of a distinct and highly visible fault, the Highland Boundary Fault, and the Southern Uplands lie to the south of another, similar line, the Southern Uplands Fault. The Midland Valley between them, although lower in altitude, is actually a graben or rift valley, and its rocks are younger than those to the north and south, including Carboniferous Coal Measures. There appears to have been considerable lateral movement along the Highland Boundary Fault as the Devonian rocks of the Midland Valley show no sign of any inclusions of (Dalradian) fragments from them until late in the period.

The Highland rocks were the most severely affected by the Caledonian orogeny. Mountains thousands of feet high, comparable with the Alps or Andes of today, were created. What is visible today in the Grampians represents the worn-down roots of these mountains, heavily metamorphosed rocks with many igneous intrusions. The Cairngorms are composed of granite from this era. Within these formations lies another tectonic feature, also highly visible, the Great Glen Fault, now containing Loch Ness. It too is Caledonian in origin with the characteristic SW-NE trend and was intermittently active until the early Tertiary. It is estimated that lateral movement along it has displaced the rocks either side of it by several hundred kilometers.

Just inland from the north-west coast of Scotland is a major unconformity, the Moine Thrust, which is best seen at Knockan Cliff, north of Ullapool. This fault plane separates the Lewisian gneiss and younger sandstones and limestones of the west from metamorphic rocks, the Moine Schist, to the east. The Moine schists, originally marine sediments, were caught up in Caledonian orogeny, and pushed westwards over the younger sedimentary rocks. The relationship between these two formations caused a major dispute amongst Victorian geologists. The problem was quite simple – the lower rocks in the sequence, unmetamorphosed sandstones and limestones containing some Cambrian fossils, were overlain by heavily metamorphosed schists, and how could that be? The famous Sir Roderick Murchison insisted that the rocks were the right way up – that they lay in the sequence in which they had been deposited. Interpreted this way, they appeared to be an extention of his own Silurian system. Objectors who read the evidence otherwise – that the rocks were in fact upside-down – were bulldozed aside, their reputations thrown into doubt. The principal critic was one Archibald Geike, later Professor of Geology at Edinburgh University. He seems not to have born the overbearing Murchison any ill-will and later wrote his biography.

It was Charles Lapworth of graptolite fame who mapped the Durness area and proposed that the rocks of the Moine Thrust were upside down. It was such an anti-establishment view that Lapworth suffered a nervous collapse. The matter was finally settled in his favour by two geologists sent out by the British Geological Survey, Peach and Horn, who published their definitive *Memoir* on the subject as late as 1907, long after the death of Murchison in 1881. Lapworth's description of the Moine Thrust was a milestone in the history of geology as it was one of the first thrust belts to be defined. Eventually, the Moine Thrust was to corroborate tectonic plate theory. The idea is that Scotland was compressed as the Baltican plate thrust northwards under the Laurentian Plate, though some North American geologists now question this, as there is a shortage of evidence for it on the ground there.

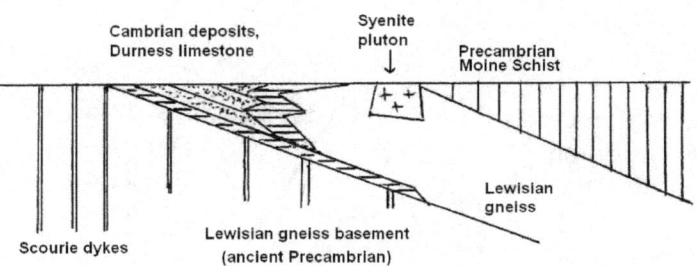

The Moine Thrust. In this section the Moine Schist is separated from the Cambrian deposits by the Lewisian Gneiss, but further south near Ullapool the older Moine Schist lies directly on top of the Cambrian deposits. This means that the section must be upside down.

****

Massive sediments of Ordovician and Silurian gritstones accumulated to be folded into the Southern Uplands of Scotland during the Caledonian orogeny. This form of compression is known as crustal shortening. At the top end of this sequence is a site a few miles west of St Abb's Head on the coast near Berwick called Siccar Point, one of the most famous sites in geology. It was here that James Hutton identified a perfect unconformity. This lay between vertical beds of Silurian slates and grits, and horizontal beds of Devonian sandstones which had been deposited on top of them. It was this structure which led Hutton to believe that the earth was very much older than 4004 BC.

Though of similar age and often of a similar type, the early Palaeozoic rocks of England, Wales and the southern part of Ireland are quite different from those of Scotland and the northern half of Ireland. The two areas lay at either side of the Iapetus ocean. The Scottish Ordovician limestones of Durness contain a marker trilobite called *Petigurus*. This small animal is also found in Newfoundland, where, as in Scotland, it was laid down in a tropical sea. This fossil is not found in other parts of the British Isles. The Iapetus Ocean which separated the two parts of the British Isles eventually closed in the Caledonian orogeny along what now forms the border between England and

Scotland. It seems amazing that this geological feature, known as the Iapetus Suture, should coincide with the modern political boundary.

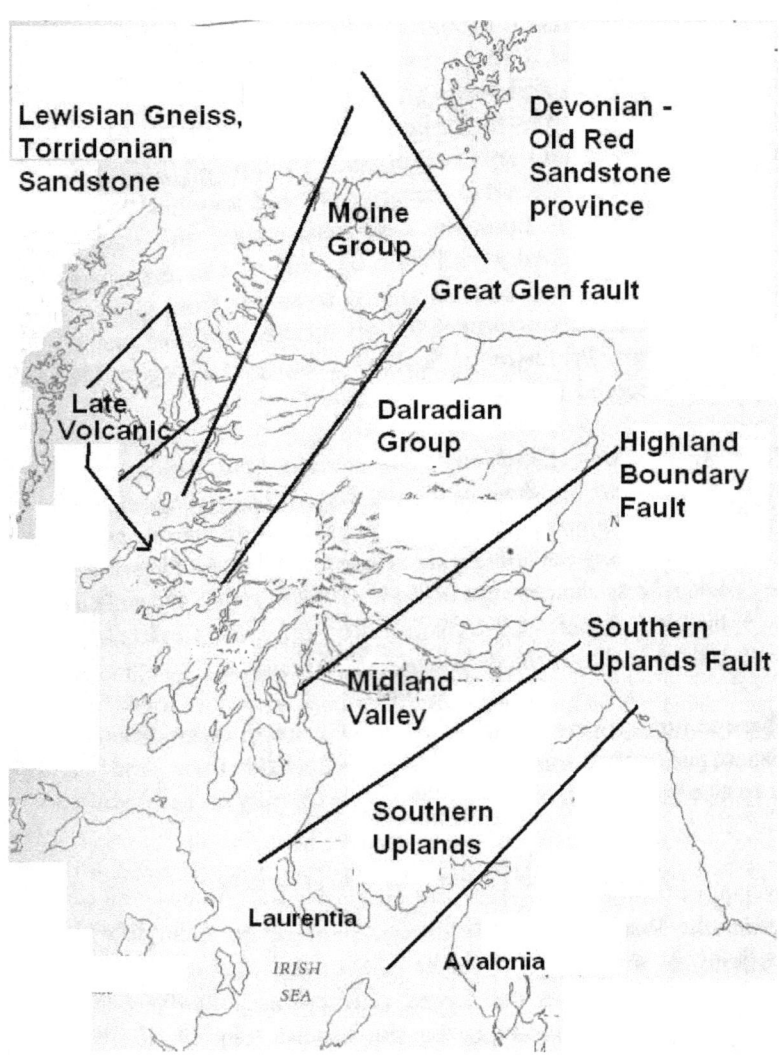

Schematic geology of Scotland. The Dalradian Group of rocks was heavily metamorphosed in the Caledonian orogeny and contains many granitic plutons such as the decorative Peterhead granite. The Southern Uplands were also folded at this time, but consist mainly of non-metamorphic greywackes. The Midland Valley is made up of Palaeozoic rocks including Coal Measures. On the islands of Skye and Mull is a Tertiary volcanic province. The far north-east, the Orkneys and also the areas at the top of the Great Glen and Siccar Point (not marked) are Devonian.

****

The geology of England is generally younger than early Palaeozoic, with the important exception of the Lake District and some Precambrian inliers. There are rocks from these periods here, but they form quite different facies to the Scottish ones, as stated above. Note the term "facies" is used in geology to indicate a distinctive rock unit, normally a sedimentary deposit, though the term can be used of metamorphic rocks.

The Lake District itself covers quite a small area but it is geologically diverse, and can be divided into different formations. The Skiddaw Group in the north contains the oldest rocks in the Lake District, Cambrian marine shales and sandstones. They have since been raised up, crumpled and squeezed, and now form relatively smooth hills dissected by deep gorges. To the south of them are found the Borrowdale Volcanics, very hard lavas and ashes formed in catastrophic eruptions and dating from the Ordovician. They make up the highest and craggiest mountains in the Lake District: Scafell, Helvellyn and the Langdale Pikes. Green Ordovician slates from this phase of Lake District geology have been widely used in the area to fashion some of the most beautiful roofs in the country.

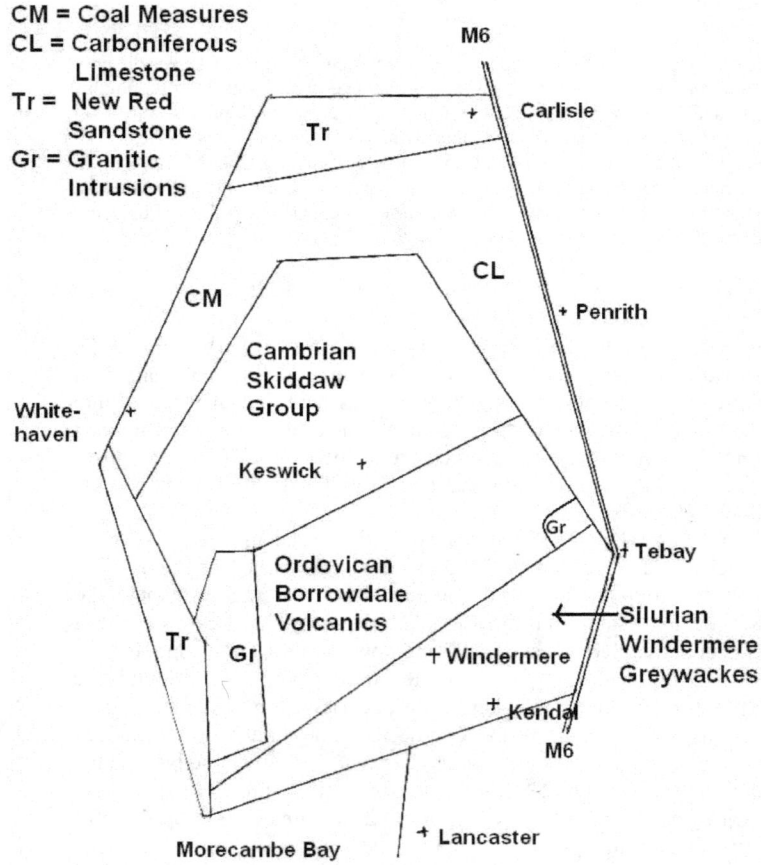

Schematic geology of the Lake District

South again is found the Windermere Group, mainly marine sedimentary clastic rocks dating from the end of the Silurian. These were later folded, faulted and eroded down to their present levels to form the lower scenery of southern Lakeland. Fossils from the lower Palaeozoic periods are very rare in the Lake District, apart from graptolites.

Next, granite plutons were intruded during the Devonian. Erosion has revealed outcrops in the south-west at Eskdale and Ennerdale, and in the south-east at Shap. These granites are thought to represent the late stages of the Caledonian orogeny and are thought to date from 390 million years ago. Finally, during the Carboniferous, a tropical sea covered the Lake District. In it was deposited the Carboniferous limestone which can be found around the eastern and southern fringes of the Lakes, although this is not a rock normally much associated with the area. During the Hercynian orogeny at the end of the Carboniferous period, the Lake District was folded into a dome structure so that the remaining Carboniferous Limestone dips away from the centre of the dome.

The Lake District as we see it today appears relatively uniform to the untrained eye. It looms up from the surrounding countryside as a large, dissected and almost round dome. It has a pleasing symmetry about it from the map, streams radiating out from the central massifs with lakes strung along them. In fact, lakes which have formed in hollows created by glaciers are considered by geomorphologists to be essentially temporary features of the landscape. Indeed, unit late in the nineteenth century, the Lake District had another lake – Kentmere, drained by the River Kent, on which Kendal stands. The exit bar from the lake was then lowered by "some humorist" (according to my old geography teacher), who blew it up.

The apparent uniformity of the Lake District scenery is not, as has been seen, reflected in the geology, except in that all the rock formations are old (Palaeozoic) and hard. Rather it is the processes which have recently acted upon these formations which have been uniform, most notably the erosion and deposition which took place in the Ice Age which ended only a few thousand years ago. It is this which has created the ridges (arêtes), corries, mountain tarns, U-shaped valleys and moraines so characteristic of the Lake District today.

****

I have spent some time myself kicking round the Lake District, and was often puzzled by the geology. I once went there when I was seventeen years old, with my schoolfriend David. I had no hiking boots – just a pair of stout shoes; no proper tent – no sewn–in ground sheet, no fly sheet to keep out the rain; and no waterproof clothing. In fact I wore a quilted anorak which just soaked up the rain.

David's parents drove us up to the Lakes and we found our way into Langdale. We erected our tent here, at the head of the valley, and at once set off to climb the Langdale Pikes, just short of 900 metres (3,000 feet) high. Even then I was curious about the geology. Many of the rocks were fine-grained and purplish in colour, but what exactly are they? Before detailed research they were dismissed as "greywackes". In fact there are greywackes in the Lakes, amongst other things, but the distinguishing feature of the rocks on the Langdales and elsewhere is their light metamorphosis. These rocks have been toasted, but only slightly. (This is indicated by the abundance of the low-grade metamorphic mineral chlorite, which imparts a greenish-blue tinge to the rocks which contain it.) We hadn't been going long before steady rain set in. By the time we got somewhere near the top, an icy torrent was pouring down, the mists had descended, and we were lost. The advice from the mountain rescue people in these circumstances is to head up rather than down, because this narrows the search area. As nobody really thinks they are going to get stranded, and as it is contrary to common sense, this advice is seldom taken. In our case we wandered around on the Pikes far too long. When we turned down the hill, I was stumbling and swearing – the first signs of hypothermia; however, we made it, exhausted, back to the camp site.

We were wet with no prospect of getting our clothes dry, but inside the tent, all was well – the sleeping bags and spare clothes were dry, so we cooked a meal in relative comfort. However, we were camped by a small stream, and at around 8.30 pm, I went outside.

"David, come and look at this!"

The babbling brook had become a raging torrent.

"What do you reckon?" said David, in consternation.

"It's bound to flood. It's still pouring down."

"You're right."

We packed up our tent then and there, hoping to find some cover indoors somewhere in the village. It seemed odd to us that nobody else on the camp site was contemplating such drastic action – in fact, nobody else did anything. We went to the pub, not knowing where we would sleep – anywhere but on the camp site. We were too young to go into the pub, so we settled down in the porch outside. This contained benches on either side, put there so that hikers could remove their muddy boots before entering the pub, and it had no door on the outside – we thought we might be able to sleep here. After a time, other campers came pouring in, wet through and in some distress.

"My car's gone!" said one man. "It floated away! The whole place is under water!"

David and I nodded to one another.

Emergency arrangements swung into place, and a caravan was found where some of us could sleep, packed like sardines on the floor in our sleeping bags. As David and I got into our bags, snug and warm, drops of water fell onto David's sleeping bag from the man whose car had floated away. He carefully brushed them off.

"Would you mind not dripping on my sleeping bag?" enquired David of this gentleman, who was practically in tears.

"Oh, sorry," he mumbled, distracted.

Foresight and planning were strong suites for David. We went home the next day, observing on our way out that the camp site looked like a lake. All the cars in it had floated away and were lined up against a farm wall. We remarked on conditions to the first person who gave us a lift, a middle-aged woman.

"This must be the worst rain in years."

"Oh no, it was just the same two years ago!"

No, not the Lake District, but scree slopes formed from basalt in the **south-west of Iceland**, another place where it pays to take your kagool

Chapter 9 – Devonian

The Devonian is the fourth geological period the Palaeozoic era spanning from the end of the Silurian Period, about 416 Million years ago, to the beginning of the Carboniferous period, about 360 Million years ago. It is named after the county of Devon, where rocks from this period were first studied. The definition of the period arose from another of those great Victorian disputes, this time between the new director of the Geological Survey, Henry De la Beche on the one hand, and our old friends Adam Sedgwick and Roderick Murchison on the other. This is known as the Great Devonian Controversy.

During the 1830s De la Beche surveyed the rocks of north Devon, where he attempted to establish the local boundary between the Lower Carboniferous and the Silurian beneath it. Here he found a layer of coal containing the remains of petrified plants, which he placed in the Carboniferous. After a field inspection, Sedgwick and Murchison said that De la Beche had made a mistake in his field mapping. All was confusion until another geologist, William Lonsdale, suggested that the fossils found beneath the coal band were similar to those of the Old Red Sandstone in Scotland. This led to a joint paper by Sedgwick and Murchison in 1839 entitled *Classification of the Older Stratified Rocks of Devonshire and Cornwall* which established a new period to come between the Silurian and the Carboniferous, and to which the fossils beneath the coal could then be assigned.

In the oceans, the first ammonites appeared, familiar from later periods in geology shops along the Jurassic coasts of England. In their curled-up form they resemble a ram's horn, and were named after the Egyptian ram-god, Ammon. However the late Devonian extinction severely affected marine life.

A common term for the Devonian is the "Age of the Fishes", as the evolution of several major groups of fish took place during the period. Jawless fishes, with simply an opening for a mouth, were the primitive early types. Now the first ray-finned and lobe-finned bony fish also appeared. All bony (vertebrate) fish have two sets of fins, pectoral and pelvic. Apart from a few oddities, all modern bony fish are ray finned, meaning that the muscles which control the fins are inside the body of the fish. In some ancient species, these muscles were grouped in lobes outside the main body of the fish. The best-known lobe-finned fish is the coelacanth, found alive and kicking in the Indian Ocean when everyone though it had been dead for 200 million years. Also appearing in the Devonian were well-equipped fish complete with jaws, which developed from the gill arches used for breathing, and also teeth.

At some point in the Devonian, what had previously been a vertebrate fish found its way onto the land and became the first terrestrial vertebrate. Recent genetic evidence showing the similarities between all terrestrial vertebrates indicates that this may only have happened once, so this humble fish became the founder of the tetrapod group which includes all amphibians, reptiles (including the dinosaurs), birds and mammals. The tetrapods ("four feet" in Greek) characteristically have four limbs, evolved from the pectoral and pelvic fins of the ancestor fish. It used to be thought that this was a lobefish, the lobes developing into muscly limbs. Modern opinion, backed by the evidence of cladistics – which takes into account all the attributes of ancestral creatures – favours the lungfish as the ancestor. Species still survive today in muddy Australian pools. After all, whatever it was would have needed lungs on the land! (In addition, the lungfish has a large bone, corresponding to the human upper arm or humerus, which attaches to its "shoulder", though in other respects its fins do not resemble the single bone – two bones – wrist – digits pattern of the later tetrapods.) This momentous event occurred around 397 million years ago. In any event – lobe or lung – modern humans have inherited some rather important features from their fishy ancestors – one central spine, two arms, two legs, jaws, teeth, lungs, and also the practice of both eating and breathing through the mouth.

Another feature of the tetrapods is five-fingeredness. Fossil creatures from the Devonian and Carboniferous have been found with different numbers of fingers, but five is the number which survived. It made the ten times tables easier! Also, as the tetrapods evolved, so did the four-limbed nature of the beast. Snakes are tetrapods which have lost their legs, and birds have used two of them to make wings instead.

One of the problems in elucidating the emergence of tetrapods from fishes has been a great shortage of actual specimens of intermediate species in the fossil record, where there was a gap of about 80 million years between the indisputable amphibians of the Carboniferous and the lung or lobefish which were their nearest known relatives from the Devonian. Over the last 20 years this position has been rectified by the full description of two early amphibians from the late Devonian. One is *Acanthostega* ("thorny covering"), an armoured newt only 7 cm (3 inches) long, which still had internal gills like a fish. The other is *Ichthyostega* ("fishy covering"), about a metre (three feet) or a metre and a half long and without internal gills. *Ichthyostega* was discovered in Greenland in the 1930s. It came to be the exclusive property of a Norwegian palaeontologist called Jarvik, under the academic rules of intellectual rights, and so beyond the reach of other paleontologists. The deadlock was eventually broken by a British scientist, Jennifer Clack, who persuaded Jarvik to allow her onto the site in Greenland to study *Acanthostega*, a much smaller amphibian found in the same late Devonian rocks. (Professor Clack, nee Agnew, started her career as a zoologist. She knew she had found her man when she met her future husband at a biker's rally and the word "Dimetrodon" cropped up in his conversation (discussed under Permian fauna below).

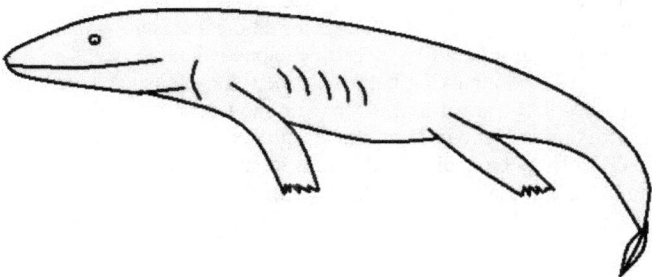

Steggy the Ichthyostega

Clack and her assistants, working on notes left by a Cambridge student from the 1970s, at first despaired of finding anything – a great risk for such costly expeditions. However specimens were eventually located and taken back to Cambridge where they were exposed after many painstaking hours – in fact months – of delicate drilling using dental equipment. This work provided Clack with the paleontologist's dream, the first complete specimens of previously barely known creatures. *Acanthostega* has both limbs and digits at the end of them – in fact eight fingers on each of its forelimbs (she had found genetic freaks, according to Jarvik, like six-fingered people). The limbs and digits are not however thought strong enough to support the animal out of water, but must have evolved to support it within water. Such animals must have later evolved the wrists and elbows needed for movement on the land. When finally fully revealed, *Ichthyostega* turned out to be a larger but similar creature, this time with seven digits on its rear limbs.

Because *Ichthyostega* already possessed fingers and toes it could tell the anatomists little about the origin of limbs. *Acanthostega* proved more interesting because of its flipper-like limbs, but scientists working in the field thought there must be other missing links, expected to date from about 375 million years ago, 10 million years before *Acanthostega*. After years of searching, the American anatomist Neil Shubin and his colleagues identified just such an earlier creature from Devonian beds on Ellesmere Island in the Canadian Arctic. Given the Inuit name *Tiktaalik*, this fish had articulated fins with primitive shoulders, elbows and wrists which would have enabled it to perform the fishy equivalent of press-ups, presumably to get out of the water and away from the large fishy predators which infested the deltas in which it lived. This animal and others like it also developed flexible necks so that they could eat without moving the rest of their body. Fishes do not have this feature – they cannot look over their shoulders. In addition, *Tiktaalik* had a flat head with eyes on the top, like the later amphibians and of course similar to crocodiles and snakes today. Fish, by contrast, have conical heads with eyes at the sides. When first announced to the world press in 2006, *Tiktaalik* hit the front pages of the New York Times.

Creepy-crawlies – arthropods – were also establishing themselves on the land in the Devonian. One of these, a giant scorpion known as *Slimonia*, nearly two metres (six feet) long, is preserved in the Natural History Museum in London.

In fact the Devonian period experienced the first significant explosion of terrestrial life – the greening of the planet and the appearance of land animals. As far as plants are concerned, it was once thought that this process began in the Devonian, but recent research has put it back into the Silurian. There is no evidence, for example, of leaves or fronds going back that far, but spores have been found, which are more easily preserved than the delicate fronds of plants. They can be isolated by dissolving sedimentary rocks in hydrofluoric acid. So it seems that the greening process took place over a time span of tens of millions of years. Nevertheless the first extensive forests covering the continents appeared in the Devonian.

One prerequisite for life on the land was the formation of the ozone layer as levels of atmospheric oxygen built up. Life in the sea such as had been lived for billions of years by the blue-green algae was protected by the surface water. The ozone layer scatters harmful ultraviolet light from the sun and so makes the earth a more habitable place. The UV light breaks up oxygen molecules (O_2) leaving spare oxygen atoms to recombine as O_3.

The higher plants of the early Devonian, the ancestors of liverworts and mosses known collectively as bryophytes, had no vascular structure at all for pumping fluids around, and did not grow much more than a few centimetres tall. By far the greatest land organism was *Prototaxites*, actually an enormous fungus which towered over other low plants. By the middle Devonian, shrub-like forests had evolved containing lycophytes, horsetails and ferns. (Lycophytes, including club mosses, differ from all other vascular plants in having leaves with only a single vein rather than the much more complex megaphylls found in ferns and seed plants.) By the late Devonian, real forests containing tall trees had appeared in the "Devonian Explosion".

One famous formation contains the remains of the Devonian flora and fauna preserved in perfect detail, the Rhynie Chert, a hard, grey rock found in Scotland in the early part of the twentieth century. Every cell was filled immediately after death by silica, so that when the chert is sliced for the microscope, the minute structure of the plants can be examined. The commonest plants are ferns and liverworts. One primitive plant preserved in the chert is *Aglaophyton*, which had neither leaves nor roots. There are animals too, spider-like creatures, mites and a very primitive insect called a springtail. These are the first known insects, about 400 million years old. Insects are arthropods with the same basic body plan as a trilobite, but they developed the ability to live in the air.

****

In Britain, by the beginning of the Devonian, the two halves of the country, north and south, once divided by Iapetus, had joined together, and indeed, enormous Caledonian mountains now towered over Wales and Scotland. Even as they are forming, however, mountains erode, and the rock formations of the Devonian represent the wastage of the Caledonian mountains fanning out onto the land and into the seas which surrounded them. All of these are still in the north and west of the British Isles. It is thought that by the end of the Devonian, the great mountains towering to 8,000 m/26,000 feet had been reduced to low hills in 40 million years, a credible rate of erosion of 1 cm every 50 years. Fragments of Dalradian rocks from the north of Scotland have been found in the Devonian sandstones of Shropshire, a distance of 500 km/300 miles!

At this time the British Isles lay roughly in the position of the Kalahari Desert and its inland Okavango delta today, in the subtropical desert belt of the southern hemisphere. Hence the most characteristic rock is the Old Red Sandstone, which is exactly what its name implies (compared to the New Red Sandstone of the later Triassic period). The red coloration of this rock derives from the abundant ferric iron which it contains. It is thought to have formed in a dry, warm semi-desert as may now be found in Colorado or Israel, subjected to flash floods and including intermontane basins and coastal plains where deposits could accumulate. As well as pure sandstone there are many formations of pebbly conglomerates and also of arkose (feldspathic sandstone). Cross-bedding – dipping bedding planes indicating the original angle of sedimentation – is strongly featured in the sandstones.

Devonian sandstones crop out in the northernmost part of Britain, in Caithness and across the Pentland Firth in Orkney, where the most famous example is the sea stack 137 metres (445 feet) high, the Old Man of Hoy (Hoy is one of the Orkney islands). Apart from the ancient Torridonian Sandstone, these rocks are the first in Britain to lie flat, in the original horizontal disposition in which they were laid down. There is another great basin of Devonian rocks around the Moray Firth in eastern Scotland, upon which the city of Inverness is built. Further south, Devonian rocks are widespread in mid- and south Wales, extending from the Brecon Beacons across the English border, notably in Herefordshire.

But of course, the place where these rocks are most prominent is Devon itself, where they are found stretching inland from both the northern and southern coasts, and form Exmoor in the north of the county. As well as the terrestrial Old Red Sandstone, there are marine sandstones. Near Torquay there are also Devonian limestones, including a rich fauna of trilobites, corals, brachiopods and early ammonites and commonly quarried for "marble" fireplaces etc. So different were the two sets of Devonian deposits, terrestrial and marine, with such different fossil assemblages, that one of the great Victorian spats took place over the differences, before it could finally be agreed that the rocks were in fact of the same age. The finer sediments of the Devonian rocks of the south-west were subsequently metamorphosed into slates during the Hercynian orogeny. These rocks are exposed on the north Devon and Cornwall coast from Tintagel southwards to Newquay.

Non-sedimentary Devonian rocks include a series of igneous rocks based around a large granite intrusion in the Cheviot Hills of Northumberland, dated to 380 million years ago. There are associated andesitic lavas and ashes. Ben Nevis, on the west coast of Scotland, has a similar combination. All these rocks have a chemical composition resembling that of the volcanoes of the modern continental margins.

****

At the end of my first long summer vacation when I was a student, then aged nineteen, I found myself in Old Red Sandstone country. I got a job working on a hop farm in Herefordshire for three weeks in September. Despite the fact that it sits on the Welsh border, and that many of the locals speak with Welsh accents, Herefordshire has the look and feel of old England. The soils, echoing the rocks, are rich and red. Most of the county is dominated by the valley of the Wye and its tributaries, winding across from England's least accessible town, Hay-on-Wye, through Hereford itself and out at Ross-on-Wye.

The farm had a small group of full-time labourers, one of whom was an old chap who spoke in the manner of Walter Gabriel. "You wouldn't be a baad-looking laad," he once told me confidentially, "if only yew'd git your 'air cut!" However most of the labour on the farm came up in a big group from the Welsh Valleys. Whole families came: the men picked the hops and the women tended the enormous hop-picking machine and picked out unwanted bits of hop bine from the conveyor belts. There must have been about 25 of them. It was towards the second half of September, but they must have been able to drop whatever they were doing and head for the farm, and they came every year.

Their leader was a giant of a man called Belly Daniels – he must have weighed 160 kilograms (25 stone). One didn't bandy pleasantries with this man. His real name was John, but he actually preferred to be called Belly. He was a mammoth drinker, and on his last day, with his money in his pocket, he drank 24 pints of cider! The Welsh crowd were overwhelming for a quiet middle–class boy like me, but one of the young women, Brenda, took at fancy to me. I could see she was eyeing me up, and she was by no means unattractive, with pale skin and ginger hair, but she was a pretty big girl, loud and abrasive. I was approached by Belly out in the fields:

"Our Brenda, see?"

"Yes, I know the one you mean."

"She fancies you, see?"
"Oh."
"She wants you to punch her brush!"
"Eh?"
"You know!" (gesture)

Nervous laugh. I think I was a bit too shy for Brenda, I never followed this up!

****

Just thinking a while about Devon and its ancient rocks, I note that this is the place where the former poet laureate Ted Hughes chose to settle. He was raised in Mytholmroyd, which can fairly be described as a grim village in the Calder Valley, between Hebden Bridge and Halifax, though there is fine scenery on the doorstep. Though he moved away from the area, first to Cambridge, eventually to Devon, Hughes left some things behind. The first of these was the body of his first wife, Sylvia Plath, who is buried in the churchyard at Heptonstall, above Hebden Bridge, an atmospheric old village formerly occupied by hand-loom weavers. Poor Sylvia put her head into a gas oven in 1963 when she discovered Hughes' affair with Assia Wevill, who subsequently suffered a similar fate; clearly, it didn't pay to get too close to our Ted. He also owned a property in Heptonstall for many years, Lumb Bank, which is still in use as a poetry writers' centre. Though he moved away himself, the Calder Valley with the moors above always remained his poetic inspiration.

Tor country on Dartmoor

Hughes lived at North Tawton, in central Devon, practically in the shadow of Dartmoor. It does not surprise me that he selected this area, because Devon is like Yorkshire with the hard edges knocked off. It is a large county with real moors, similar to the Yorkshire moors, but in the case of Dartmoor, with a sea view! I went there to explore for myself in 2010. It is like Yorkshire in other ways – though in the south of England, most of it is still north of the Tees-Exe line, and it is equally remote from London. It has little of the metropolitan chic of Cornwall, and there is not so much work so the property prices are sensible. Also it faces the wet winds from the Atlantic on hills rising to over 600 metres (2000 feet) right on the coast, so though not as wet as the Lake District or Snowdonia, where the mountains are higher, I wouldn't forget to take your umbrella if I were you.

Dartmoor is a granitic inclusion dating from the Triassic, as are Bodmin Moor and the massif of Land's End. Dartmoor is famous for its tors, extended rocky outcrops at the summits, which again find echoes in the Millstone Grits of Yorkshire. The other Devon moor, Exmoor, is quite different – it is composed of Devonian sandstone and has no tors.

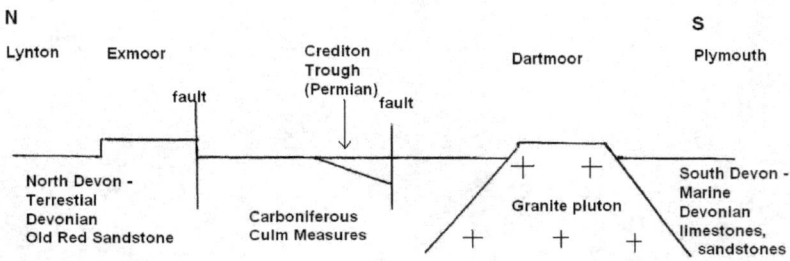

Cross-section of the geology of central Devon

****

We have now encountered all four of the great disputes which caused such a stir amongst the geologists of the day, but can you remember what they were? The most important by far was the furore surrounding Darwin and his *Origins of the Species*, containing as it and its successor volumes did the implication that the earth is almost immeasurably old, much older than indicated by the Old Testament, and effectively writing religion out of geology. The second was the argument about the boundary between the Cambrian and Silurian systems which took place between Sedgwick and Murchison, eventually resolved with when a whole new period, the Ordovician, was carved out between them by Charles Lapworth. The third also involved Murchison and his Silurian System, and concerned what came to be known as the Moine Thrust in Scotland, where older rocks lie on top of newer ones, a fact denied by Murchison but once again asserted by the later Charles Lapworth. Finally there came the great Devonian debate, where the marine rocks in the south of the county were eventually put in the same period as the entirely different terrestrial sandstones of north Devon.

Of all these controversies, the one which has lasted the longest has been the definition of the top of the Cambrian. In Sedgwick's original type area, north Wales, his worst fears have now been realised, not by Murchison's Silurian, but by Lapworth's Ordovician. The regions now universally considered to be composed of Cambrian rocks are really quite small, the Llanberis slate belt lying to the north of the Snowdon massif, and the Harlech Dome to the south of it. Modern geologists as late as the 1980s have revisited and remapped the area, and as a result, the Ordovician has been extended. (This proves that modern geologists do not spend all their time peering down electron microscopes!)

.

Chapter 10 – Lower Carboniferous

The Carboniferous is the geological period which stretches from the end of the Devonian Period, about 359 million years ago through to about 300 Million years ago. Its name is derived from the Latin for coal, *carbo*; Carboniferous means "coal-bearing". Many coal beds were laid down globally during this time, hence the name. However, the earlier part of the period is dominated by the deposition of massive limestones and then (millstone) grits. This distinction is recognised in North America where the Carboniferous is treated as two geological periods, the earlier Mississippian and the later Pennsylvanian. In economic and scenic terms, the Carboniferous is by far the most important geological period of any encountered so far in the British Isles.

The main early Carboniferous plants were ferns, horse-tails and club mosses, plus a number of the first gymnosperm groups. The term "gymnosperm" comes from the Greek word meaning "naked seeds", because the seeds have no casing. This contrasts with the seeds of flowering plants, the later and now-dominant angiosperms, which are enclosed. By far the largest living group of gymnosperms are the conifers (pines, cypresses etc). Typical of this flora is the monkey puzzle tree (*Araucaria araucana*), a pine with an ancient lineage and appearance. One look at the vicious leaves of this tree is enough to understand why it has persisted for so long.

Terrestrial life was already well established by the Carboniferous period. Amphibians were the dominant land vertebrates and indeed the period is sometimes called the Age of the Amphibians. An amphibian is still very much a water animal, and lays its eggs in water. Living in the air meant important physical changes for these creatures. They changed the way they walked, so instead of bending their bodies from side to side, like a fish, and leaving two limbs on the ground, they left three legs on the ground while walking, raising their bodies in the air. Again to assist with walking in this way, articulated elbows, knees, wrists and ankles developed. Their limbs, however, still stuck out of the sides of their bodies, instead of growing directly underneath (in other words like a newt and unlike a cow). Fins, gills and fishy scales disappeared as these species adjusted to life on the land.

Meanwhile on the sea bed there were forests of crinoids, a type of sea-lily or echinoid. Whole limestone cliffs came to be composed of their skeletons. When the *Challenger* set out once more to explore the ocean floor in the early part of the twentieth century, these creatures were dredged up from the depths in large quantities, to the astonishment of the scientists on board. In addition, the Carboniferous was the heyday of the brachiopods, some of which grew very large in size.

One marine predator has been with us ever since Devonian times, and flourished in the Carboniferous – the shark. Many species within this group have been engineered with marvellous precision by nature to be predators. The shark was once thought to be an evolutionary throwback, as it has a backbone made of cartilage rather than bone. As some sharks do grow bony parts, it is now realised that this is just another means of adding flexibility to this most dangerous animal. A shark is not intelligent, in the way that say a tiger is. It will attack instinctively, ruthlessly and inexorably and cannot be deflected by any guile or reason. In fact sharks do not fossilize well because of their soft skeletons, but plenty of their teeth have been found. These themselves are evolutionary adaptations, sloping backwards in the mouth to foil the escape of prey.

A global drop in sea level which had taken place at the end of the Devonian was reversed early in the Carboniferous. This created widespread shallow seas around the edge of the continents where carbonate deposition could take place. There was also a drop in south polar temperatures. The southern part of Gondwana was glaciated from the middle Carboniferous, 330 million years ago. These conditions scarcely affected the deep tropics, where warm swamp forests flourished within 30 degrees of the northernmost glaciers.

****

The Carboniferous or Great Limestone is the most widespread rock formation of all in the British Isles – a king amongst rocks, upon which some of the country's finest scenery has developed, instantly recognisable to all. The Mendip Hills consist of Carboniferous limestone, showing notable geomorphological features, including Cheddar Gorge and the Avon Gorge near Bristol. Outcrops occur around the edge of the coalfields in south and north Wales, where the Great Orme near Llandudno is an outstanding feature. There are a few outcrops in Shropshire such as Titterstone Clee Hill and at Little Wenlock. It covers much of the area of the Derbyshire Peak District.

However, the main outcrop in terms of area is in the Pennines and the south and east of the Lake District. It is also found in the Midland Valley of Scotland, and is the single most common surface rock in Ireland, where it outcrops from the east to the west coasts. In some places the rock is heavily folded, especially at the Great Orme.

Carboniferous limestone is a sedimentary rock made of calcium carbonate. It is a tough rock, generally light-grey in colour. It was formed in warm, shallow tropical seas teeming with life. The rock is made up of the shells and hard parts of millions of sea creatures, some up to 30 cm (12 inches) in length, which became encased in carbonate mud. Fossil corals, brachiopods and crinoids are very much in evidence as components of Carboniferous limestone; indeed the rock is full of fossils.

Carboniferous limestone is extensively quarried for many purposes. It is crushed for roadstone and aggregate wherever it outcrops, particularly in the Mendips, the Yorkshire Dales and north Wales. In certain places, as at Horton in Ribblesdale in the Pennines, it is sufficiently pure for the production of chemical-grade lime. It is used to make flue gases for industry, and for cement manufacture. In many places it is metalliferous, and has yielded lead (in the Peak District and Weardale), and copper (in North Wales, where important Bronze Age mines are to be found inside the Great Orme). Indeed the Yorkshire Dales are full of the remains of lead mining.

The Carboniferous limestone is far from being pure limestone, as it is normally deposited as part of a sequence – first the pure limestone, then shale, then sandstone (which may terminate in a fossil soil or seat earth), then limestone again. Such sequences, known as cyclothems, represent the various inundations and regressions of the sea. The limestone forms when the sea is the deepest, and the sandstone at its shallowest. Cyclothems are found in other formations including the succeeding Millstone Grit and Coal Measures, but in the Carboniferous limestones they are known as Yoredale cyclothems, after the River Ure (Yore) in north Yorkshire. The defining character of these Carboniferous cyclothems, all of which were laid down on or near the coast, is given firstly by limestone, then Millstone Grit – a grit being a coarse sandstone – and finally coal. The sea level changes which created them may have involved local subsidence, worldwide changes due to fluctuating ice sheets at the south pole, or simply changes in the local deltas as occur on the lower Mississippi or Yellow Rivers today, when the big river can simply shift its mouth by several hundred kilometers. The cycles can be very numerous – up to a hundred in any formation.

Malham Cove, Yorkshire, an ice-age waterfall composed of Carboniferous limestone

Only recently I had cause to venture into the limestone country of the Yorkshire Dales, approaching from the north. The Lake District must have formed an island in the Carboniferous seas, around which corals grew. Hence the Carboniferous limestone is to be found immediately to the east of the Lake District and is possible to drive from Kendal to Sedburgh – a distance of only 20 km (12 miles), across the

M6 – passing from one national park into another, and from the dark crags of the Lake District to the smoother white stone hills of Dentdale. The family of my great-grandmother Nelly come from here, but moved to Bradford in the 1850s. In my hand that day was a copy of a letter written in 1889 by my great-grandfather Peter, still in Bradford, to his young wife. She had taken my grandmother and her sister Lotty off the visit her uncle at a place called Peas Gill House in Dentdale. He asked her to bring back a supply of eggs and butter. Her access must have been via the Settle to Carlisle railway, which passes the head of Dentdale at a high altitude to the east.

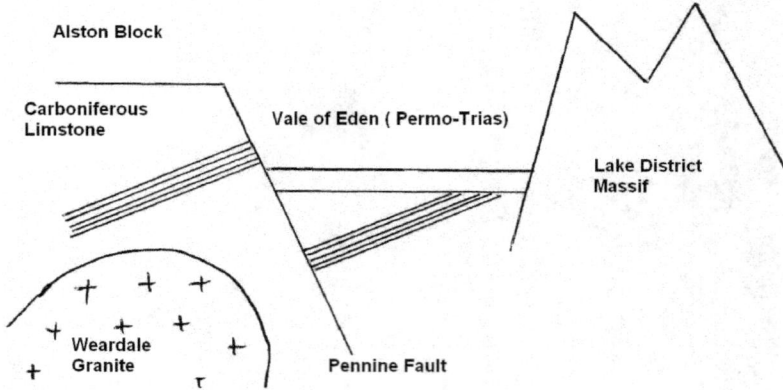

The Alston Block of the north Pennines is a Carboniferous structure and a good example of a normal fault, here known as the Pennine Fault. The higher rocks of the Alston Block (upthrow, eastern side) are buoyed up by the Weardale granite, which is a relatively lightweight rock. On the downthrow (western) side is the Vale of Eden, where the Carboniferous rocks are overlain by a lower but later strata of Permo-Triassic age. To the west of that looms the Lake District.

I found Peas Gill House, now mostly a ruin, on the side of the hill where a minor road takes off southwards through a steep-sided valley called Barbondale. Adam Sedgwick himself sorted out the geology of this area, for a major feature known as the Dent fault passes through Barbondale, separating the Silurian rocks of the Howgill Fells of the Lake District to the west from the limestones and interbedded sandstones to the east. A considerable vertical movement is thought to have taken place along this fault – thousands of feet – so that the Silurian rocks lie much higher than the Carboniferous.

At the farm there was a stream tinkling through the adjacent field, lambs gamboling; the sun shone down of the green fields and higher moors of the fells. What a pleasant change this must have made from the family home near the cemetery in Bradford, where my great-grandfather worked as a stonemason. Judged from the average age of death at the time, he must have been a busy man. His own family was not to escape the scourge of the age – tuberculosis. My poor great-grandmother Nelly – a very pretty woman, from the two photographs we have – died of the disease when she was only 33 years old, leaving my grandmother a child of nine and her sister Lotty only five. Lotty too was destined to live a short life, dying unmarried at the age of 31. My great-grandfather Peter only lasted until he was 51. At one time he had tried to make a living in the United States, and set off alone to do so, settling briefly in Denver. However he plainly missed his family – we still have his letters – and went home to Bradford. Looking out across the beautiful fells of Dentdale that day in 2011, it felt like my own personal version of the TV program "Who do you think you are?" – and very sad for all that.

The Yorkshire Dales are not the last conspicuous feature of the lower Carboniferous. The same limestone outcrops on the Northumbrian coast at Holy Island, perhaps better known as Lindisfarne, an island just off the coast which has a long history. There is a monastery here, which looks like a castle, and which was once the home of St Cuthbert, patron saint of old Northumbria. The ancient Angles were once besieged here by the Briton, Urien, only to be relieved when Urien was murdered by one of his fellow chieftains. This was also the site of the first Viking raid on Britain, in 793, and so the starting place of a whole new phase of English history.

Also from this period are Arthur's Seat and Castle Rock in Edinburgh, volcanic plugs – the solidified vents of old volcanoes – both of these adding to the impressive gothic gloom of this atmospheric city. It was at Salisbury Crags, a dolerite sill which lies beneath Arthur's Seat, that James Hutton demonstrated the intrusive nature of igneous rocks as they worked their way into pre-existing local sediments. (This rock has been extensively quarried to build Edinburgh's old town.) The Midland Valley of central Scotland is in fact a rift valley, bounded by the Highland Boundary and Southern Uplands faults. Arthur's seat was a rift valley volcano, similar to Kilimanjaro in the African Rift valley today. The Midland Valley also contains Coal Measures from the lower Carboniferous, the oldest in northern Europe. Also from this era is the dolerite sill on which Stirling Castle is built.

Finally in Northumbria once again is the Whin Sill, an enormous, flat intrusion of igneous dolerite into the Carboniferous limestone, 73 metres (237 feet) deep at it thickest. It must hence postdate the limestone and is dated to the late Carboniferous. Once again, it is a rock with a famous history, as Hadrian's Wall was built on top of it where it forms steep north-facing cliffs. It underlies much of north-eastern England and reappears again in Teesdale, where a famous waterfall, High Force, tumbles over it. It outcrops offshore on the Farne Islands and Lindisfarne.   There is no other sign of contemporary volcanic activity in this region. Dolerite when injected is relatively mobile and the Whin Sill dolerite is believed to have spread from the area of Arthur's Seat.

Hadrian's Wall winds its way on top of the Whin Sill

I went myself to Hadrian's Wall with a small party from school when I was seventeen, accompanied by a single Classics master. That trip is rememberable for two things. Firstly, I forgot to pack any eating utensils – a mess tin, knife, fork, spoon and mug. This may not have mattered, but the previous summer I had taken my A-levels at the young age of seventeen and had obtained the best results in the school. So I had a reputation as a brainy bonce. This omission of the eating utensils did not however look particularly clever, and Fearnley, a boy a year younger than me, was quick to take advantage. He exclaimed "what an idiot!" every time mealtimes came round. After a couple of days I was quite ready to ram Fearnley's knife and fork down his throat for him, and see how he liked that! The second thing was that it was the October half-term and it was dark and wet on the Wall. The going was tough and everyone carried a heavy pack. Walking with packs is more difficult by an order of magnitude than walking without them. At the end of the first day of crawling up and down – the country on which the Wall stands is anything but flat – all day until dark, we consulted our maps, confident we had covered fifteen miles. We had not – we had only gone eight miles. Still we had seen the most spectacular section of the wall at Crag Lough near Hexham, which owes its scenery to the Whin Sill, which then stretches away for miles to the south under the billowing folds in the landscape. Nevertheless after three days and two nights of constant drenching, by common consent, we headed for home.

****

Between the Carboniferous Limestone and the Coal Measures lie the Millstone Grits. Geologists all over the northern hemisphere can subdivide the epoch of the Millstone Grit (known as the Namurian) by the evolution of marine fossil shells called goniatites (which resemble ammonites). The rock formations are divided into sixty zones over a period of 12 million years – that is, each zone represents a period which on average is 200,000 years long. This is a very fine definition of geological time for a period as old as the Carboniferous.

Like the Great Limestone, the Millstone Grits were deposited in cyclothems. The grits themselves are coarse, hard sandstones but they are interbedded with much softer shales and more sandy, finer-grained sandstones or flagstones. Roadside cuttings frequently show a vertical series of beds. The grits frequently show signs of cross (current) bedding; this is characteristic of deltaic deposits.

The Millstone Grits are very prominent in the central Pennines where they stretch from the northern part of the Peak District into Airedale in the Yorkshire Dales and then across to Morecambe Bay, taking in important towns including Halifax, most of Harrogate, Blackburn and Lancaster. They form high ground of bleak moors – in fact the word "bleak" could have been invented to describe this scenery, covered in heather, cotton grass and bilberries. This is the land of the Bronte sisters, and Wuthering Heights corresponds to a real place, Top Withens, near Haworth where the Brontes lived. The grits also form the highest mountains in the Peak District, including Kinder Scout – but the atmosphere is archetypally northern: grey, hard and bleak. Into the Yorkshire grits are cut steep valleys including the Colne and the Calder, where the terraced hillsides are dotted with ancient and pretty yeomen's houses dating back to the seventeenth century. This area has a long history in the woollen industry and was as heavily populated as it is now back in the time of Daniel Defoe in 1724.

The hard, impermeable rocks proved ideal for the cutting of millstones, used to grind corn. Also there are occasional spectacularly-shaped outcrops carved by erosion. The rocks are a favorite for climbers and there are many well-known sites for this activity, as at Stanage Edge to the west of Sheffield. Some sandy facies of the grits, notably the Elland Flags, have another import function, as they make the best paving stones in the country. The streets of London are not paved with gold, but with Elland Flags and similar facies. Large parts of the local towns are also built of the stone. Incidentally, professional geologists very much dislike the use of the term "Yorkstone" to describe the Elland Flags, because the same term is used by stonemasons to describe many other types of stone which happen to outcrop in West Yorkshire.

An important system of faults known as the Craven Faults separates the Great Limestone (north) from the Millstone Grit (south) sequences. These three faults run from the west of Ingleton down to the region of Grassington (north) and Gargrave (south), a distance of 50 km/30 miles or more, roughly marking the south-western boundary of the Yorkshire Dales national park. To the south of the faults lies the Aire Gap, through which the River Aire flows on its way from its source at Malham Cove to Leeds.

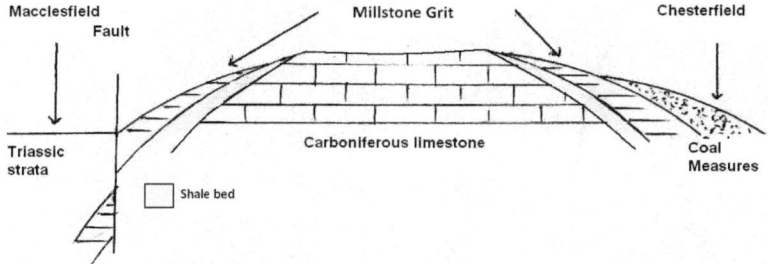

Diagram to illustrate the geology of the Derbyshire High Peak District. It is a denuded anticline with the oldest rocks, the Carboniferous Limestone, exposed in the centre. The younger Millstone Grit and Coal Measures would once have covered the whole area. On the western side a fault forms the boundary with the Triassic rocks of Cheshire which outcrop at lower elevations than the older rocks to the east. To the north in West Yorkshire, the Millstone Grit cap remains, but north again, in the Yorkshire Dales, it has once more been eroded away.

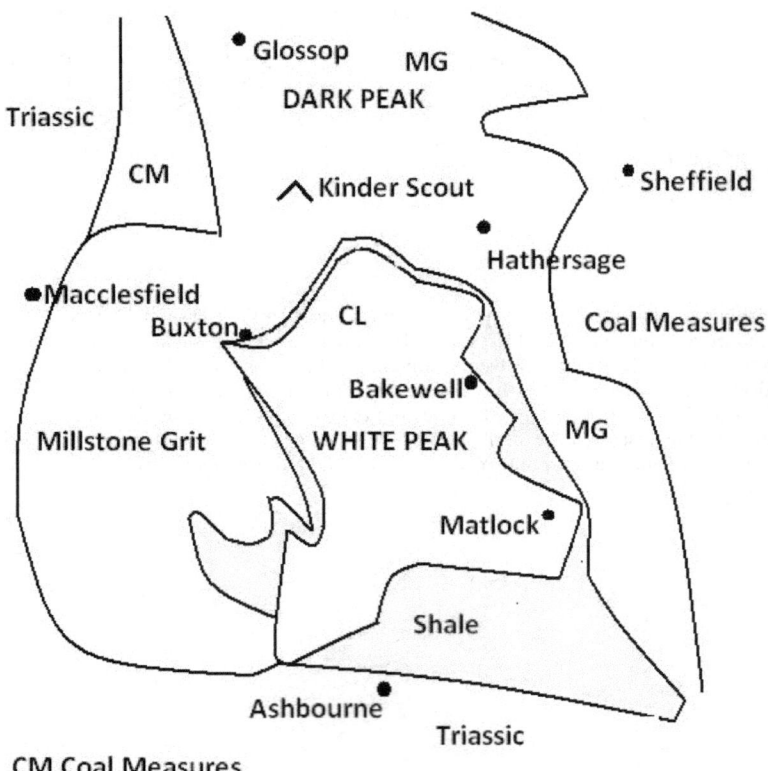

CM Coal Measures
MG Millstone Grit
☐ Carboniferous Shale
CL Carboniferous Limestone

Schematic map of the geology of the Peak District. The limestone White Peak area is relatively small, but it attracts a lot of visitors.

****

Several important canals were built though the central Pennines, despite the high ground, including the Rochdale and the Huddersfield Narrow Canals. On both the Yorkshire and Lancashire sides of them, the Coal

Measures outcrop, and there were coal mines at one time on the eastern side of Halifax and Huddersfield and the western side of Burnley and Rochdale. One of the main functions of the canals was to be to move coal from the Coal Measures area into the Millstone Grit area. The geology of the area is of supreme importance, because the rainfall runs off the Millstone Grits in streams of some of the softest, sweetest, most beautiful water in the country, ideal in fact for use in the manufacture of textiles. To the north in the Yorkshire Dales, and to the south in the Peak District of Derbyshire, the Millstone Grit gives way to limestone. This dissolves to harden the water. The economy which grew up in these areas is completely different – there are no mill towns.

Chapter 11 – Upper Carboniferous

The Coal Measures are the next rocks in the Carboniferous system above the Great Limestone and the Millstone Grit. The large coal deposits of the upper Carboniferous primarily owe their existence to two factors. The first of these is the evolution of lignin, which made possible the growth of tall trees by reinforcing the cell walls during growth. Tall trees themselves are an engineering marvel, vast hydraulic systems for raising water from ground to canopy by capillary action. It is thought that these mechanisms developed quite quickly in the simple competition for light. The second factor facilitating coal formation is the lower sea levels that occurred during the late Carboniferous as compared to the Devonian period. This allowed for the development of extensive lowland swamps and forests in North America and Europe. The late Carboniferous also experienced periods of glaciation, and latterly, mountain building.

It is thought that large quantities of wood were buried during this period because fungi and bacteria had not yet evolved which could digest lignin effectively. Eventually – after fifty million years – fungi evolved which could do this. During this long period, a massive carbon sink was created. Woody plants took carbon dioxide from the air, as plants always do, and released the oxygen. The carbon however was not returned, but remained in the trees in life and in death. This had the knock-on effect of increasing the amount of free oxygen in the atmosphere, as there was less carbon available to form carbon dioxide. It is estimated that the proportion of oxygen in the air increased from 20% to 25%, and it was this which allowed for the evolution of the giant millipedes and dragonflies so characteristic of the period. *Meganeura*, a giant dragonfly-like creature, was far larger than any comparable creature today. A dragonfly as big as a seagull was found in the Coal Measures at Bolsover in Derbyshire, the first beast of Bolsover (the second was its MP, Dennis Skinner) – it was the largest flying insect that ever lived.

The Coal Measures represent the remains of deltaic sediment, and consists mainly of clastic rocks (shales, siltstones, sandstones and conglomerates) interstratified with the beds of coal. The coal seams themselves are generally about 30 cm (one foot) thick and occupy only a small proportion of the total depth. Fossil soils or seat earths often occur within the sequence. These are known as fireclay if they are clayey, or ganister if they are sandy. In both cases the plants have abstracted the alkalis from the soil. As the alkalis act as fluxing material, this means that these rocks are now very resistant to heat, and so they are used to make bricks and ceramics which need to resist high temperatures. The ganister is used directly to make hearth bricks.

The coal itself is derived from the remains of the rich Carboniferous forest, the best known tree in which is the lycopod *Lepidodendron*, a tree with a characteristic lozenge-shaped bark pattern. Lycopods formed great trees at this time, resembling palms in appearance, though they are unrelated. Their remains were preserved in anaerobic conditions, that is, without oxygen. From time to time marine inundations occurred which drowned the forest and left its remains at the bottom of the sea, under a layer of sand and silt, eventually to form coal seams. The thin marine sedimentary bands contain ammonite fossils which can be used to give relative dates to these inundations. These bands are found in every geographical area of the Coal Measures – for example in Britain, the United States, Germany and Poland. Their distribution has revealed a telling fact – the marine bands are the same age everywhere. The only explanation for this can be that the inundation of the coal swamps was part of a worldwide and significant rise in sea level, such as would flood the modern equivalent areas today – for example Bangla Desh or the Amazon delta. This in turn can only realistically have been caused by fluctuations in the size of the Carboniferous ice sheets – it is thought that the south pole was glaciated from the middle of the period – and so by changes in atmospheric temperatures (and so chemistry). This obviously resonates loud and clear with modern forecasters of global warming – it has happened before, and look what happened then!

In Britain, the Coal Measures are widespread north of the Tees-Exe line. Notable outcrops occur either side of the Pennines in Yorkshire and Lancashire. The fact that the Coal Measures lie either side of the older and higher-lying Millstone Grit in the central Pennines tells us something about the geological structure, for this area is a denuded anticline. The hills form an arch with its axis running north-south, and the central beds raised above the outer beds. At one time the Coal Measures would have covered the whole area, but have been eroded away on the top of the hill, exposing the Millstone Grit. Other Coal Measure are found in Nottinghamshire and elsewhere in the Midlands; on the coast of Cumbria; in Northumberland and Durham; in the Midland Valley of Scotland; and in South Wales. In this last place there is a limited surface outcrop as the Coal measures lie in a syncline running east-west across Glamorgan, largely beneath a later

Carboniferous formation, the Pennant sandstones. The area is deeply dissected by the Rhondda and other valleys and the coal is anthracite, which has the highest proportion of carbon of any form of coal. This is because the sulphur and other impurities have been squeezed out of it by light metamorphism. It lies in narrow, broken seams and has proved a valuable resource, if difficult to win.

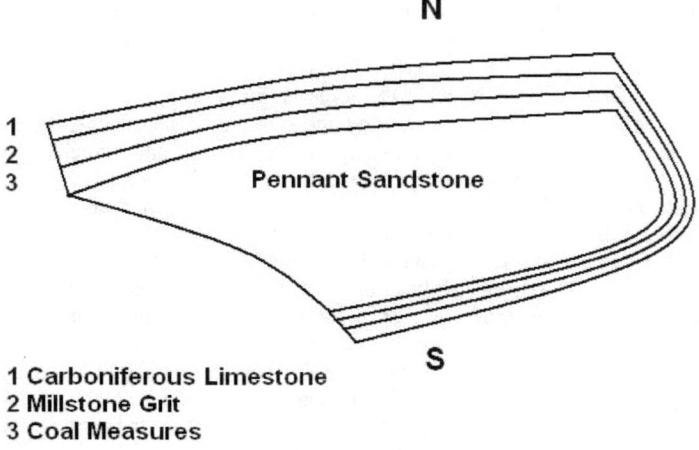

1 Carboniferous Limestone
2 Millstone Grit
3 Coal Measures

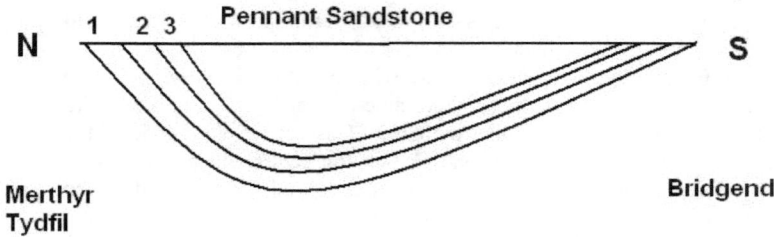

The South Wales coalfield. This is a syncline or downfold, where the rocks are progressively younger from the outside inwards. The top diagram shows the area as it looks from above, as portrayed on a simplified geological map. Synclines often look like this, with the newest, most central rocks having the largest surface exposure, even though the outcrop may be thinner than the older rocks below it. The second diagram shows a cross-section of the syncline.

****

Coal measures are also found, and have been mined, under younger rocks in Kent and Somerset, which, when you come to think about it, are funny places for coalfields. They also extend under the North Sea, lying beneath sandstones of the Permo-Trias capped by evaporites (evaporated salt deposits). The gases contained in these coal measures were not squeezed out by the Hercynian orogeny, as they were in South Wales, and have risen from the coal up to the level of the evaporites, which are impermeable to them. These ancient gases provide much of the modern North Sea gas.

Coal measures are also found in Shropshire, where they include sills, horizontal intrusions of basalt or dolerite. One of these, found today at the top of Titterstone Clee, is used to make beautiful stone sets, used for street paving in towns such as Ludlow.

The coal measures of Scotland include bands of oil shale. In the 1850s a certain James "Paraffin" Young developed a means of extracting crude oil from these rocks. This has had a notable effect on the landscape, as the area west of Edinburgh is dotted by the most enormous red slag heaps or "bings" which have been left behind.

There are also extensive contemporary deposits known as the Culm Measures in north and central Devon, but they do not contain coal.

In the eastern United States the term Coal Measures is applied to the Pennsylvanian coalfields, where they outcrop in the valleys of the Allegheny and Monongahela Rivers (the two rivers meet at Pittsburg).

****

The Carboniferous coal beds provided much of the fuel for power generation during the Industrial Revolution, which maps to them quite closely, and are still of great economic importance. From the middle of the eighteenth century, one of the main coal mining areas of England was centred around the town of Wigan in Lancashire, and it was to distribute coal from Wigan that a whole network of canals was constructed. Over the summers of 2002, 2003 and 2004, I sent off on my bike to explore these canals. This story is told in my book *Tyke on a Bike*. However, I was not the first person to become curious about Wigan. George Orwell spent two months in the "industrial areas" of the north of England in 1936, and the result was his famous book, *The Road to Wigan Pier*. This is his description of conditions for the working classes in general and the coal miners in particular during the Great

Depression. He concentrated on mining areas, particularly Wigan and also Barnsley and Sheffield, and it is a dismal picture that he paints of Wigan, showing vividly the impact of coal mining on the landscape and on the community.

"Labyrinthine slums and dark back kitchens with sickly, ageing people crawling round and round them like black beetles....monstrous scenery of slag-heaps, chimneys, piled scrap-iron, foul canals, paths of cindery mud criss-crossed by the print of clogs...belching chimneys, blast furnaces and gasometers......The houses are poky and ugly, and insanitary and comfortless...they are distributed in filthy slums round belching foundries and stinking canals and slag-heaps that deluge them with sulphurous smoke."

This type of housing was for the lucky ones – the unlucky people had to live in "caravans". Orwell writes: "Along the banks of Wigan's miry canal are patches of waste ground on which caravans have been dumped...The majority are old single-decker buses which have been taken off their wheels and propped up…"

Nor was Orwell under any illusions about the social character of such places:

"In a Lancashire cotton-town you could probably go for months on end without once hearing an 'educated' accent, whereas there can hardly be a town in the South of England where you could throw a brick without hitting the niece of a bishop."

Was it still like that in Wigan? I finally got there, riding all the way through the Yorkshire Dales on my bike along the Leeds and Liverpool Canal. Despite the unprepossessing air of the suburbs of Wigan today, it is very clear that things have moved on a long way from Orwell's time. I had heard that Wigan town centre was surprisingly good, following the implementation of a planning decision in the 1920s requiring the construction of black and white timber fronts to new buildings. So I strolled up into the town, and it was all true. It was amazingly pleasant and prosperous-looking, and the black and white timber fronts were indeed there. There was an arcade, modern I think, but blending in nicely, a good church, and a number of decent buildings squashed into a limited area. There was also a large, new Debenhams store. It was all so much better than say Rochdale or Blackburn.

Moving away from Wigan and back towards Leigh and Manchester, there is a dramatic change of scenery immediately after the town. The countryside is dominated by large lakes or "flashes" either side of the canal, caused by mining subsidence. Back in the eighteenth and nineteenth centuries, this area was central to the economy of the whole north-west. Today it looks as if it has barely recovered from its centuries of exploitation. Though some scars have healed, much of it has an air of desolation. There isn't even much in the way of agriculture along parts of this stretch of canal, just weeds. There are slag heaps (used for that most irritating of sports, moto-cross) and quite a number of bridge footings, the bridges themselves, presumably for railways, long gone.

Having said that, the largest of the flashes (one outside Wigan and one outside Leigh) have been landscaped with trees. They were dotted with the white sails of dinghies, moving quickly in the breeze, and the prospect across them was fair, suitable for nice houses in fact. (The sort of nice house that ceases to be nice when you wake up in the morning 20 feet underground.) The waterfowl and seagulls were also making the most of this wetlands environment. Hey, it's the Lancashire Broads! (The Norfolk Broads are also man-made, dug out for peat.) If only George Orwell could see this, he would be amazed. Housing subsidence is one of the themes of his book.

Also in *Tyke on a Bike*, I rode my bike along the Union Canal, which runs westwards from Edinburgh. Here I had my first encounter with a "bing":

"Most striking from the towpath was an enormous slag heap which towers over the town of Broxburn, straight ahead. This looked for all the world like a red and menacing volcano, devoid of vegetation as if recently erupted. The closer I got to this thing, the more monumental it looked. Cut into fantastic shapes by erosion, it must be 200 feet high at least. Known locally as a shale bing, it represents the remains of mining for oil shale. If ever it became unstable Broxburn would be engulfed, like Abervan. Beyond Broxburn there were several more, not quite as large."

"I resumed at Winchburgh on another evening, heading west for Linlithgow. Passing along a long wooded cutting another great slagheap hove into view. After this point the view became very scenic - the monstrosity was behind me. Visible from the towpath were the tops of the Forth rail and road bridges, and a huge full moon hanging just over the horizon – could it be the harvest moon? I passed a pair of joggers, a man and a woman, puffing and blowing a bit, while I glided along in high ratio. Yes, I thought, this biking beats jogging any day."

****

One more Carboniferous formation lying immediately above the productive Coal Measures in the English Midlands is the Etruria marl, used to make the original Staffordshire pottery. Marls are a friable, earthy mixture of clay and calcium and magnesium carbonates, often full of shells. The Etruria marl was suitable for the manufacture of the large, rough pots which were made in Staffordshire before the invention of china. Josiah Wedgwood (1730-95) was one of the founding fathers of the Potteries and he called his house Etruria. Bone china, made in England from the 1790s, is much finer than the old rough pots and requires the use of china clay (kaolin is the Chinese word for this), which had to be imported from Cornwall. Here it is found around the margins of the great granites where they have been rotted by erosion. More canals were needed here to get the china clay in, and the pottery out in an unbroken state. Josiah Wedgewood was the grandfather of Charles Darwin, and it was his fortune that enabled Darwin to settle down in Kent to his leisurely if not entirely unstressed life as the first-ever popular science writer.

****

The Hercynian (or Variscan) orogeny is a mountain-building event caused by Late Palaeozoic continental collision between Euramerica and Gondwana to form the supercontinent of Pangaea. It is known as the Alleghanian orogeny in North America, where it threw up the Ouachita ("Wichita") Mountains of Arkansas. In the same time frame, much of present Siberian plate welded itself to Europe (Baltica) along the line of the Ural mountains, then much higher than they stand today. Some of the largest copper mines in the world today are related to ophiolites along the line of this closure. Most of the Mesozoic supercontinent of Pangaea was now assembled.

The British Hercynian belt includes the mountains of southwestern Ireland (Munster), Cornwall, Devon, Pembrokeshire, the Gower Peninsula and the Vale of Glamorgan. The trend of the mountains is east-west, in contrast to the Caledonian SW-NE trend. Hercynian folding is also present in France in Brittany, the Ardennes, the Massif Central, the Vosges and in Corsica. Particularly famous features from the Massif Central are the puys, small hills. Those in the north in the Chaîne des Puys, including the famous Puy de Dome, are in fact cinder cones of recent origin. The plugs rising from the town of Le Puy-en-Velay in the south are original Hercynian volcanic features.

Volcanic plug at **Le Puy-en-Velay**

In Germany the orogeny created the Black Forest and Harz Mountains. Its presence is felt in evidence in the middle Rhine Valley, which is a "graben", a tectonically created ditch bounded by faults from the higher "horst" on either side of it, which constitutes the Eifel, Hunsrück, Taunus and Westerwald massifs. In its time the orogeny created huge ranges, the rumps of which remain with us today at much lower relief.

Just as the earlier Caledonian orogeny of the Silurian set the scene for the creation of vast sandstone beds created by downwash from the mountains in the Devonian, so the Hercynian mountains created the same conditions for the formation of the New Red Sandstone of the Permo-Trias.

The later part of the Carboniferous witnessed the onset of the Permo-Carboniferous Glaciation. This was the first real ice age since the end of the Ordovician. The sequestration of large quantities of carbon as Coal Measures may have helped to bring it about, by reducing the amount of carbon dioxide in the atmosphere and so reducing its effect as a greenhouse gas. The glaciation was intensified because a major land mass, Gondwana, lay over the south pole, allowing ice to accumulate over millennia, as it does now over Antarctica. The presence of glacial till in now widely-separated continents was one piece of evidence brought forward to support the existence of Gondwana, and so of continental drift. Known as tillite, this rock was first identified below Permian sandstone in the Great Karoo of South Africa, and was subsequently recognised on other Gondwana sites as far north as Oman in the Arabian Peninsula.

****

So, have you ever gone out and bought a geological map on the scale of 1:50,000? In the case of British maps, this is where you move away from the comforting generalizations of the small-scale (1:625,000) maps of northern and southern Britain. Fine maps in their way, these still present the national geology very much in the manner of William Smith, for example with broad swathes of familiar Jurassic and Cretaceous strata marching across the map in bold colours. When you finally decide to get out of your armchair, however, and do some real field geology, then you need a 1:50,000 map. Suddenly all the comforting certainties dissolve into a mass of detail – hundreds of different rock types on a single sheet, faults everywhere, surface drift, alluvium and landfill sites obscuring the bedrock, deeply incised valleys exposing dozens of facies – all this invokes an immediate sense of confusion to the unwary beginner, if not outright panic.

However, all is not lost. As a matter of fact, the best place to start is not the surface map of the horizontal layout of the strata, but the vertical profile which will be found below it. This shows the rocks as they were deposited or intruded. I have in front of me now the 1:50,000 sheet for Huddersfield, which at first appearances is bewilderingly complex. However the vertical profile shows that, though riven by many faults, the stratification is relatively simple. The is a "Caledonian Basement" approximately 2600 metres (8500 feet) down; above that about 1400 metres (4500 feet) of Lower Carboniferous (including the Great Limestone); above that, the Millstone Grit series, about 900 metres (3000 feet) thick when buried at the surface; and above that at the surface towards the east, the Coal Measures. Each of these is given the name of the appropriate geological epoch: Lower Carboniferous, Dinantian; Millstone Grit series, Namurian; and Coal Measures, Westphalian. Reaching to heights of about 460 metres (1500 feet) in the east at the apex of the Pennines, the Millstone Grits dip gently to the east, so that at Halifax 25 km (15 miles) from the top of the Pennines, they dip underneath the Coal Measures. These in fact rear up in a steep escarpment above the town on the eastern side. However anyone standing on the Pennine Moors should realise that only 500 metres (1600 feet) beneath his feet, if that, lies the Carboniferous limestone which he will find all around him if he moves 50 km (thirty miles) to the north-west, or 65 km (forty miles) to the south.

The problem then is to get used to the idea of a cyclothem as it is found on the ground, because these Namurian and Westphalian rocks were all laid down in a great delta area where the shoreline advanced and receded repeatedly. The Millstone Grits include truly massive sandstones which were indeed used to make millstones; one such is shown on the map as the "Rough Rock". However the Coal Measures which lie immediately above the Rough Rock shows no such simplicity. Layers of mudstones (shales) alternate with hard, thinner bands of sandstone flags and bands of coal. Each band of coal is identified on the map, be it only a few inches thick (the thickest on this map is 1.8 metres or nearly six feet). As part of my work for the West Yorkshire Geological Trust I inspect exposures of rock as they are found in road cuttings and quarries. The Millstone Grit exposures do tend to exhibit fairly massive sequences, but the Coal Measures invariably show a rapid alternation, including the famous fossiliferous marine beds.

The **chateau at Polignac**, near Le Puy-en-Velay, built to take advantage of a volcanic plug left over from the Hercynian

Chapter 12 – Permo-Trias

The Permian is the geological period which extends from 299 to 251 million years ago. It was named after the Perm region of Russia by our old friend Sir Roderick Murchison in 1841. The Czar of Russia, Alexander II, was so delighted with Murchison's work in his country that he gave the geologist a gold snuffbox, studded with large diamonds, which only goes to show that, then as now, geology can pay!

The world at the time was becoming a very uncomfortable place, hot and dry, and dominated by Pangaea. The extensive rainforests of the Carboniferous had shrunk, leaving behind vast swathes of desert. The very fact that there was only one large continent led to a loss of diversity in the flora and fauna, as successful species could spread over huge areas and eliminate ecological competitors that would have been safe on separate continents. Also the fact that there was only one continent meant that quite a lot of it must have been in rain shadow, as the Gobi Desert is today – far away from ocean winds, which have already shed their moisture on windward peaks. Deserts seem to have been widespread on Pangaea.

Permian deposits around the world are not as widespread as might be expected, but are especially notable in three regions: the Ural Mountains (where Perm itself is located), China, and the southwest of North America, where the Permian Basin in Texas is so named because it has one of the thickest deposits of Permian rocks in the world.

There are two sets of zonal fossils (used for correlation between different regions) for the Permian, ammonites and fusulinids. These latter were single-celled floating creatures which formed part of the foraminifera. On the land, the great swamp-dwelling trees, lycopods such as *Lepidodendron* and *Sigillaria*, were gradually replaced by early conifers and seed ferns. The most important ecosystem here is known as the *Glossopteris* flora, which dominated the southern part of Pangaea. In the air, giant predatory dragonflies (*Odonata*) buzzed about.

In terms of land vertebrates, the Permian saw the emergence of true reptiles from the previous amphibian forms. Reptiles no longer had any need to lay eggs in water. Instead the water was trapped inside the egg! Unlike amphibian eggs, reptile eggs have hard shells and contain amniotic fluid to sustain the embryo. Also, instead of being fertilised externally by sperm ejected into water, these eggs also came to be fertilised internally – the reptiles had invented sex!

Towards the end of the Permian, the archosaurs appeared – ancestors of the later dinosaurs. What actually emerges in the Permian is the first full terrestrial food chain, complete with large herbivores and carnivores. Though the players within it would change completely, the structure would remain until modern times. This pattern had already emerged with other ecologies – for example coral reefs and forest.

The best-known Permian predator is a sail-backed reptile called *Dimetrodon*, often called a Permian dinosaur, though it is actually a synapsid, a reptile with one set of holes either side of its skull. Reptiles and their successors were at first incapable of jaw movements – for chewing and so on – but as the complicated musculature developed, so did holes in the head. Tortoises and turtles have no holes. Diapsids have two pairs, and evolved into dinosaurs, birds and crocodiles. The synapsids also included a fearsome predator, *Anteosaurus*, which again, to the average museum visitor, is a very good imitation of a dinosaur, though it predated true dinosaurs by millions of years. It is classed as a therapsid reptile, an advanced group of synapsids which dominated the late Permian and also known as the mammal-like reptiles.

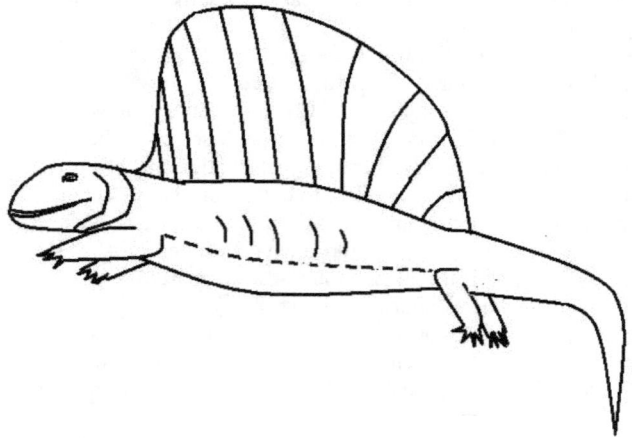

Dino the Dimetrodon

The top part of the skull of *Dimetrodon* showing the cavity behind the eye socket, which is found in all synapsid reptiles.

The therapsids had virtually replaced *Dimetrodon*-type creatures by 240 million years ago. They had abandoned the sprawling newt-type limbs in favour of legs and feet which tucked directly under the body, allowing for faster movement. To accompany this change, the bottom set of ribs disappeared, as they got in the way of the lower legs, and a stronger lumbar (lower back) region developed to support the stronger back legs and spine. Elbows and knees became fully articulated, the upper limbs pointing backwards at the elbow, and the lower limbs pointing forward at the knee. All mammals and bird retain this orientation, though it is disguised in some creatures (dogs and horses). Also appearing at the end of the Permian were the first cynodonts ("dog teeth"), therapsids which would go on to evolve into mammals during the Triassic.

The Permian ended with the most extensive mass extinction event recorded in palaeontology. 90% of marine species became extinct, as well as 70% of all land animals (according to Peter Toghill, 95% of both). It is also the only known mass extinction of insects. Trilobites were never seen again, nor were fusulinids, nor Palaeozoic coral. The massive expansion of life which had continued from the end of the Ordovician came to a shuddering stop, bringing an end to the entire Palaeozoic era. However it was not a sudden process as might have been brought about by a catastrophe – it was spread out over several million years.

There is significant evidence that massive basalt eruptions lasting thousands of years in what is now the Siberian Traps contributed to environmental stress leading to the mass extinction. (The term "trap" is derived from the Swedish *trapp*, step or stair, to describe enormous continental basalt lava flows which weather into stepped plateaus.) The reduced coastal habitat and highly increased aridity probably also contributed. Based on the amount of lava estimated to have been produced during this period, the worst-case scenario is an expulsion of enough carbon dioxide from the eruptions to raise world temperatures five degrees Celsius. It can hardly be a coincidence that there was also the greatest marine regression of Phanerozoic times at this crucial time, when the sea withdrew from the continental shelves, a fact which also means that there is a also shortage of marine deposits in which to hunt for evidence of the extinction. Because so much of the mantle had erupted on top of the continental crust in Siberia, it is understandable that sea levels should fall as mantle material from beneath the sea bed moved to fill the gap left by the eruptions. Note also that this same set of circumstances – the formation of huge "traps" (in the Deccan of India) and a sea level regression – is found in the mass extinction at the end of the later Cretaceous period.

However even the eruption of the Siberian Traps would not increase world temperatures enough to explain the death of 90% of marine life. One idea is that this warming could have slowly raised ocean temperatures, causing the variation in temperature between the poles and the equator to diminish, so that oceanic currents became sluggish. The oceans began to stagnate, until frozen methane reservoirs below the ocean floor near coastlines melted. These methane stores, known as methane clathrates, methane hydrate or methane ice exist today. According to the theory, this release would expel enough methane into the atmosphere to raise world temperatures an additional five degrees Celsius. (Note, however, that methane is not stable in the atmosphere of the earth, and soon changes into carbon dioxide and water vapour.)

In any event, the earth began to roast. Most of the species on the planet then died of suffocation – from a lack of oxygen. There is evidence for this in the iron pyrites (fool's gold) crystals found in late Permian sediments, as this only forms in conditions of oxygen starvation on the sea floor. Modern research has shown that the level of oxygen in the atmosphere dropped from 21% 280 million years ago to 15% by 260 million years ago, then to only 10% at the end of the Permian 251 million years ago. It was a bad time to be around.

Instead of invoking non-uniformitarian ideas (such as the release of buried methane), it might be best to seek a solution for Permian conditions on the earth as we see it today. Sixty percent of the continental area has agglomerated into one "World Island", as Halford Mackinder called it – Europe, Asia, Africa, Arabia and India form a single land mass. The vast majority of the people living in that land mass live in Europe, India and south-eastern Asia, including eastern China. The rest of it is very lightly populated because it is too cold and/or more especially too dry. Obviously there are large deserts in Africa and Arabia. Central Asia is very dry, with several bone-dry deserts, the Karakum, Kyzyl Kum, Takla Makan and the Gobi, surrounded by a much greater area of semi-arid steppe which stretches from north of the Caspian Sea all the way to Mongolia. Ulan Bator has rainfall of only 200 mm (8 inches) a year. Northern Siberia may not look like a desert, but that is because transpiration and insolation are so low. It receives very little precipitation. Even the teeming millions of India and China only live in favoured parts of those countries, relatively small areas compared to the whole. China has a vast hinterland of dry areas, including Tibet, Sinkiang (Chinese Turkestan) and most of northern China. There are always new plans to try to put more water into the northern Yellow River. Whole countries – Egypt, Iraq, Pakistan – rely on water brought to them on rivers from mountainous areas beyond their borders. The huge area stretching from the south-west of Iran through Baluchistan and the Quetta district of Pakistan and on to the Thar desert of Rajasthan is dry and supports very few people from its own water. So even today, a large proportion of the World Island suffers from rain shadow or lies under an area of convective subsidence in the atmosphere. Also, the Himalayas act to block out wet tropical maritime air from Central Asia, as they run west to east across the path of it. North America does not have this problem – the Rockies run north-south and warm, wet air from the Gulf of Mexico can penetrate all the way into Canada.

In other periods, it can be shown that the earth is favoured when warm equatorial waters can circulated easily around the continents, and that when they cannot, disaster results. Thus it was that the balmy Eocene descended into the ice ages of the Pleistocene. Therefore it seems clear that rather than invoking methane clathrates, it is simpler to look at the configuration of the continents, and if the ocean currents cannot circulate through the continents, and if those continents are not riven by oceans which can water their coasts, there will be a problem, and there was indeed a problem in the Permian.

This aridity was however a feature of most of the Permian and the next period, the Triassic. Life must have been very stressed in the middle of it, at the Permo-Triassic boundary. The eruption of the Siberian Traps might have been enough to finish a lot of it off.

There is recent evidence that the Permian mass extinction was caused by a major meteor impact. Geochemists have identified fullerenes, which may have come from outer space (where these molecules have been identified) or have been created by the impact. They are spherical molecules of carbon ("buckyballs" – in fact resembling soccer balls) containing 60-200 carbon atoms. These contain trapped helium and argon where the isotopes do not correspond with those normally found on earth, though they do match the isotopic signatures of these elements coming from meteors. There is also evidence from China that the earth was enveloped by a sulphurous cloud, and the further possibility of layers of glassy microspherules such as may have been created by a meteor impact. So this isn't very strong evidence, but more of it may yet be found.

In Britain there are very few Permian outcrops apart from some sandstones and conglomerates in south Devon and the Magnesian Limestone of eastern England. As far as Britain is concerned, the Permian is usually included with the succeeding period, the Triassic, as the Permo-Trias, of which there are substantial outcrops. The British Isles lay at this time in roughly the climatic position of the Persian Gulf today – a desert with warm offshore seas, which were sometimes cut off from the main ocean and so forming evaporites, deep beds of salts evaporated from the sea.

****

The Triassic is the geological period that extends from about 250 to 200 million years ago. It is the first period of the Mesozoic Era. Both the start and end of it are marked by major extinction events. It began in the wake of the Permian mass extinction, which left the Earth's biota severely thinned out. Recovery from this extinction was protracted, and is thought to have taken 20 million years for terrestrial creatures, 30 million for marine life. Pangaea existed until the mid-Triassic, after which it began gradually to split into two separate landmasses, Laurasia to the north and a second Gondwana to the south. In between them lay the "Tethys" ocean, named by the Austrian geologist Eduard Suess after the wife of the god Oceanus. It persisted for millions of years – the Mediterranean Sea is the rump of it still remaining. Suess himself was a very influential man who wrote a famous book, *The Face of the Earth (Das Antlitz der Erde)*, and we shall meet him again.

Pangaea reached its maximum extent at the start of the Triassic, and included all modern continental areas apart from China and part of south-east Asia. The global climate in the Triassic was mostly hot and dry, perfect conditions for the formation of red sandstones and evaporites, but the earth became cooler and wetter as Pangaea then drifted apart. There is no evidence of glaciation at or near either pole throughout the period.

The end of the Triassic was marked by another mass extinction in which approximately 80% of species disappeared. The last epoch of the Triassic, known as the Rhaetic or Rhaetian, saw continued coal formation in China.

On land, conifers (gymnosperms) flourished in the northern hemisphere. *Glossopteris* (a seed fern with a leaf shaped like a tongue, its name based on the Greek word for "tongue") was the dominant southern hemisphere tree during the early Triassic period. It was part of a sub-polar flora, the southern equivalent of the coniferous forests of Russia today. In fact the distribution of the *Glossopteris* flora, typical of cool climates, was one of the factors which led to early speculation about continental drift, as there was a common flora across the constituent parts of Gondwana. It was first identified in Permian sandstones in South Africa, then in conglomerates in India, but is unknown in the northern hemisphere. This name – Gondwana – taken from the name of a tribe in India, the Gonds – was coined at the end of the nineteenth century by Eduard Suess, based on evidence such as came from *Glossopteris*. It means Land of the Gonds so the "land" in the usage Gondwanaland is superfluous.

The mapping of the ocean floor has shown that there is more to old Gondwana than Suess could ever have known. For example, the continental shelf around Australia extends northwards to Papua New Guinea and southwards to Tasmania. Lying to the east of Australia are two considerable undersea plateaux, rifted away from the main block of continental crust of Australia, but not very far away, and which are composed of similar rocks; also around Australia are several smaller "islands".

In the Triassic seas, new types of coral appeared to replace the species wiped out in the Permian extinction. The ammonites recovered, diversifying from as few as five remaining species that had survived. For the first time, marine reptiles began to appear in numbers, evolving from terrestrial ancestors and going back into the sea which their earlier forbears had left a hundred million years earlier. These included the first plesiosaurs (long neck and tail, four paddles) and the highly successful ichthyosaurs (dolphin-like), which appeared in early Triassic seas and soon diversified, some eventually developing to a large size (ten metres, 33 feet) during the late Triassic.

Pterosaurs, flying reptiles, first appeared during the Triassic, already fully formed as flying creatures. Their lifestyle was unusual as they required all four limbs to fly, so could not run on the ground. Eventually they evolved into large, frequently bizarre and evidently terrifying forms. Also appearing at this time were the first turtles and crocodiles. Archosaurs (primitive dinosaurs) were initially rarer than the therapsids which had dominated Permian terrestrial ecosystems, but they began to displace them in the mid-Triassic.

The first known real dinosaur, the *Eoraptor* from Argentina, is dated to 228 million years ago, the middle of the Triassic. Here was a new type of animal, built for speed and agility. It had strong but light bones and a distinctive hip structure which gave it an upright, bipedal posture. This was made possible by a dinosaur innovation – a specialised, articulated ankle. A similar animal was *Coelophysis*, hundreds of skeletons of which have been dug out of a place called the Ghost Ranch in New Mexico. Within a few million years, different types – heavy, four-limbed sauropods ("lizard foot") – had appeared. The dinosaurs did not immediately become dominant, but much of the competition was wiped out by a mass extinction at the end of the Triassic. There is speculation that this could have been caused by a meteor strike in Quebec.

During the late Triassic, advanced cynodonts gave rise to the first mammals. The cynodonts form a perfect link between reptiles and mammals. There is some indication that they were developing warm-bloodedness, and they also had eardrums. They also had specialized teeth – molars, incisors and canines – this is a distinctive mammalian feature; and they may have had hair, as skulls shows small holes in the snout thought to carry nerves to whiskers. These physical changes were to be the mammal's legacy from the reptiles, together with waterproof skins without scales, and the development of hard-shelled or leathery eggs which could be hatched away from water.

So how is a mammal to be distinguished from a therapsid like a cynodont? There are big differences. Mammals all have hair or fur, or evolved from furry ancestors. The females have milk-producing mammary glands, and all are warm-blooded, but that in fact only gets us as far as a monotreme. There are three kinds of mammal – monotremes, marsupials and placentals. The monotremes ("one hole") seem likely to have come first, because they still laid eggs, which when hatched were then suckled on their mother's milk. Monotremes have only one hole out of their body, so the intestine, bladder and reproductive tract all exit at the same place. There are very few monotremes left on the planet today: the most famous is the Australian duck-billed platypus, an animal thought to be a hoax when first described. Another is the echidna, also from Australia, which resembles a hedgehog. Marsupials and placentals both give birth to live young, so have passed the egg-laying stage and lost the need to build nests. The marsupial young are born very small and are then nurtured externally from their mother's milk in a pouch. Placental babies grow much larger internally in their mother's bodies, nurtured by the placenta. Having reached mammalian stage, however, there were still many changes needed to make a person from a small furry creature, but there was time enough to do it – another two hundred million years or so.

****

For most of the Permian and Triassic, the British Isles lay at latitudes 20 – 25 degrees north, exactly where the Sahara Desert is found today, and it had a Sahara-type climate including prevailing easterly winds, also just as today. It was also in the middle of the continent of Pangaea, so marine formations are comparatively rare. The rocks are most terrestrial formations deposited as dunes, by major rivers originating outside the country or in wadis, or by flash-flooding and other such semidesert processes.

Permo-Triassic rocks are widespread throughout England outcropping in a broad band of SW-NE trend – the trend of most of the geology from this period onwards – literally from the Exe to the Tees. This stretches from Cornwall and Devon to Somerset and the mouth of the Severn, and on through the east and west Midlands before branching either side of the Pennines to Cheshire in the west, and to Yorkshire in the east, where it reaches the coast at the Tees estuary. Along the western flank of the main Triassic outcrops, from Nottingham through York to Sunderland, is a Permian rock, the Magnesian Limestone, reaching its widest outcrop on the coast at Hartlepool. This a limestone made of magnesium and calcium carbonates. It makes an attractive creamy-yellow building stone. Unlike the later New Red Sandstone, barren as one might expect of a desert rock, the Magnesian Limestone is full of fossils – clams, brachiopods, sea mats and so on. This formation is now classified as part of the Zechstein group, named after the ancient Zechstein Sea and formed in a basin which stretched from eastern

England across Europe into Poland. Now known as the Cadeby formation, the rock was used as the original building stone of York Minster, one of the finest cathedrals in all England, and also of the Houses of Parliament. However, it has limited ability to withstand chemical weathering, especially if salt is involved, and other limestones have been tried at York Minster during periodic renewals of the building. The stone used to replace magnesian limestone in the Houses of Parliament is inferior oolitic limestone from Rutland.

The Midlands region is largely based on Triassic rocks – Worcestershire, Staffordshire, the Birmingham area, Warwickshire, Nottinghamshire, Derbyshire, Cheshire and Leicestershire included. Wales and the Pennines were uplifted along faults relative to the basins between them, which the above areas represent. This movement was of the order of two kilometers or a mile and a half vertically.

North again, there is another Permo-Triassic basin centred on the Solway Firth and Carlisle, stretching across to Penrith and Appleby. Triassic sandstones form the red cliffs at St. Bees Head in Cumbria. The landscape is mostly broad valleys covered with rich red soils. At the lower end of the sequence lies the early Triassic Bunter Sandstone (also called the Bunter Pebble Beds), a pebbly alluvial (i.e. formed by rivers) rock popular for use in monumental masonry – including whole castles – at many sites in northern Europe. It tends to form well-drained but infertile soils and is the rock which underlies Cannock Chase in Staffordshire, an Area of Outstanding Natural Beauty (this designation often means that the land is not much use for farming!) It is thought to have been deposited by a major Nile-style river running northwards across southern Britain. The dominant Triassic rock however is the New Red Sandstone, the outwash from the Hercynian mountains. Given that formed in this way, it is by no means a pure sandstone, as it contains a quantity of feldspar. This is the building stone used for Liverpool Anglican Cathedral.

A further facies from the late Triassic is the Keuper Marl, siltstones and mudstones found in the Midlands, Devon and north Yorkshire, now renamed the Mercia Mudstone.

There is another other important Permo-Triassic facies, also connected with the presence of sea water. This is the evaporites. These are salts which came out of solution in sequence – first, the hydrated form of calcium sulphate known as gypsum (plaster of Paris), then rock salt (sodium chloride), then anhydrite (the anhydrous sulphate of calcium), then salts of potassium and magnesium. The rock salt of Cheshire was formed by the evaporation of the Triassic seas. These salts were exploited in the Middle Ages for use in preserving food, and went on to form the basis of modern chemical industries in Cheshire (including in the Wirral). Other Triassic salt deposits are found around Teesside which later gave rise to the chemical industries of Billingham. Given that they are evaporites, these rocks are enormously thick, up to 1000 m/3250 feet, and unlike modern evaporites in that they contain a lot of potassium. If the Mediterranean was to dry out today it would yield only about 25 m/80 feet of halite (sodium chloride salt). The evaporites of the north-east are thought to have formed from the Permian Zechstein Sea, which repeatedly dried out and was then replenished from the east.

There are deep potash mines in North Yorkshire to this day (potash is a term for various salts of potassium, including potassium hydroxide). The potash is used for making fertilizer amongst other things. These various salts are also used to make plasterboard, ammonia, sulphuric acid and hydrochloric acid.

The desert rocks of the Permo-Triassic were laid down when Britain lay in the subtropical desert belt of the northern hemisphere, in roughly the position of North Africa today. Hence there is a striking correspondence with the Devonian Old Red Sandstone, which was laid down when Britain was in the desert belt of the southern hemisphere. Between these periods came the Carboniferous, when Britain had drifted across the equator, with its coral reefs and coal swamps.

****

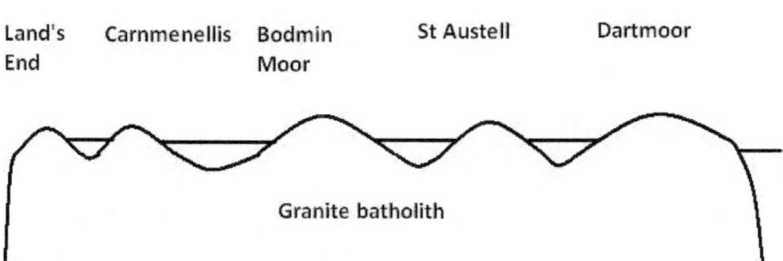

The granite domes of Cornwall and Devon are all part of a single structure, most of it underground. What looks like the infill between them is not infill but older rocks from the Devonian, into which the granite was intruded millions of years later.

Some of the most important rocks of the Permo-Trias are not sedimentary at all, but igneous – the great granite domes of Dartmoor, St Austell, Bodmin Moor, Carnmenellis, Land's End and the Scilly Isles in the south-west. These intrusions form the tops of a great underground mass of granite or batholith. The granite cooled slowly at great depths (in the case of Dartmoor thought to be 17.5 kilometres or ten or twelve miles down, gradually rising up from that level)) and was exhumed at the surface only much later. It is a very tough rock, lacking the bedding planes which can easily be exploited by weathering, though it is jointed. The famous tors of Dartmoor are the tops of the outcrop, showing the effect of weathering on the jointed structure. Around the edges of the granites is a zone of metamorphosis known as an aureole, in places as much as seven km (four miles) wide. The granite itself forms visible veins in the country rock, leaving no doubt that it is the younger in age and flowed as a liquid.

These enormous, vegetation-free cliffs form part of the **Pink Granite Coast** of Corsica.

The origin of these and other granites was the subject of much debate in the middle years of the twentieth century, known as the Granite Controversy. Arthur Holmes and his second wife, Doris Reynolds, a rare thing – a female geologist – thought that existing schists and gneisses were infused with a hot "emanation" which transformed them *in situ* into granites, possibly without any magma stage. Their group was known as the granitisers. The opposing party, the magmatists, led by Canadian Norman Bowen, thought that the granite had solidified from molten gneiss at depth in the crust. A problem for the granitisers was the loss of calcium, iron and magnesium which would have been necessary to turn basic gneiss into acidic granite. Experiments with high-pressure ovens under the supervision of Bowen eventually showed that as temperature and pressure rise, a mixture of gneiss and water turns itself into a granitic magma. This magma has a lighter density than gneiss and so rises through the crust. The energy to heat the crust is supplied by subduction as one tectonic plate dives under another, throwing up nappes and new mountains and eventually exposing the tops of batholiths. Arthur Holmes may have let his heart rule his head in this dispute and his reputation suffered late in his career for it.

The granites of the south-west have had a massive impact on the history of the whole country, because they are riven by mineral veins which contain a most useful metal – tin. The British Isles were known to the ancient world as the "Tin Islands" (Cassiterides – said by some to be the Scilly Isles, where there is no tin!). The metal was an essential requirement in the production of bronze (where it is combined 10/90 with copper), the metal at the foundation of the whole Bronze Age. Copper was also mined in commercial quantities in Devon and Cornwall, but it is a much more common metal than tin. If there had been no tin, would there ever have been a Roman invasion – would we have Hadrian's Wall? The Romans conquered England for supposedly political reasons, but then, the same could be said of the British behaviour in the Boer War, which was really fought over the gold mines of South Africa. The tin and copper mines of Cornwall and Devon were a massive part of the British economy well into the twentieth century and have left an unexpected legacy for the unwary tourist, a landscape of slag heaps and pit villages! Employment in the tin and copper mines reached its zenith in the middle of the nineteenth century, before the opening of new mines overseas made the mines of the south-west unprofitable. However, the miners found that their skills were in great demand in the new mines, for example in northern Michigan, but also in many other places, including South Africa, Australia and Chile, so many emigrated, taking their families with them. It is estimated that half a million people left the mining communities, the greatest numbers emigrating in the 1870s.

Where the granite has weathered deeply, it has left a clay containing the mineral kaolinite, used in the manufacture of fine china, and so another important industrial resource. Superheated water from hydrothermal vents is also thought to have contributed to the kaolin supply, rotting away the feldspars from within the granite. The extraction process for this has left another set of slag heaps. The modern economy has had to establish itself in this scenery. One aspect of this is the Eden Project, a series of large greenhouses on the site of a former claypit just outside St Austell, the largest town in Cornwall and possessing the unlovely aspect of a mining town.

Chapter 13 – Jurassic

The Jurassic is the geological period which extends from about 199 to 146 million years ago, that is, from the end of the Triassic to the beginning of the Cretaceous. The Jurassic constitutes the middle of the three periods of the Mesozoic era, also known as the age of dinosaurs. The term "Jurassic" is taken from the Jura Mountains of the Franco-Swiss border. The most famous and typical Jurassic rocks of all, however, are found on the south coast of England, especially in Dorset. The start of the period is marked by the major extinction which took place at the end of the Triassic. However, the end of the Jurassic period did not witness a mass extinction. As far as the British Isles are concerned, the desert period of the Permo-Trias was over as the warm Tethys Ocean spread over the land to lay down marine rocks full of ammonites.

There are lots of Jurassic formations in western Europe, as much of the continent lay submerged under shallow tropical seas. These deposits include two famous sites where soft body parts have been preserved (*lagerstätten*), Holzmaden and Solnhofen in Germany. The first *Archaeopteryx* – one of the world's most famous fossils – came from Solnhofen. It bears a strong resemblance to the small carnivorous dinosaur *Velociraptor*, and has a full set of teeth, but is feathered, and so must be a bird. The astronomer Fred Hoyle thought it was a fake, like Piltdown man! It is one of the few examples of a true missing link found in palaeontology – in fact the absence of missing links has allowed crazy Creationists to claim that the Lord created each species from scratch. About ten million years older at 160 million years, another dinosaur bird has recently been found in Liaoning, China, named the *Anchiornis* – the skeleton and feathers of which prove conclusively that feathered birds evolved from small dinosaurs.

In North America, the best-known Jurassic rocks are not marine, but alluvial, laid down by rivers. Such is the Morrison Formation, the biggest dinosaur graveyard in the world, of which more in due course.

Sea levels were high in the Jurassic, at times at least 100 metres (325 feet) higher than they are now. The causes of marine transgressions such as this are examined in the next chapter. The flooded lowlands provided friendly environments of shallow, warm, seas round the edges of the continents ("epicontinental"). Within these seas, fishes underwent the "Mesozoic Marine Revolution" with the appearance of the teleost group. These fishes had jointed jaws which gave them a great evolutionary advantage over the existing sharks and bony, heavily-scaled sturgeon-like fish. Marine reptiles included ichthyosaurs, which were at the peak of their diversity, plesiosaurs and pliosaurs. One of the plesiosaurs was *Liopleurodon*, an enormous predator with vicious teeth, thought to grow to 150 tonnes in weight.

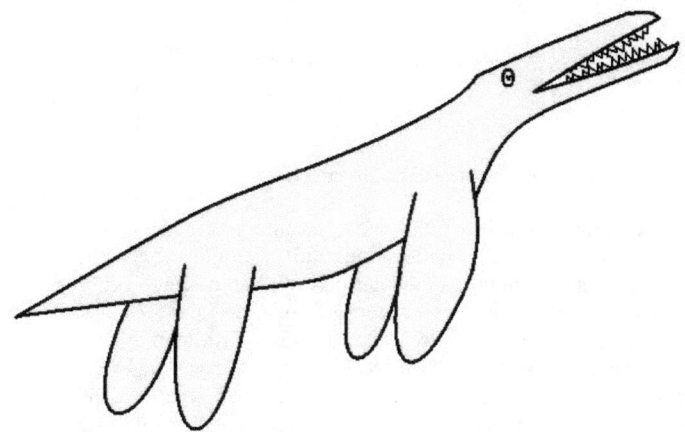

Leo the Liopleurodon. Note the short, stubby tail, large flippers, short neck and long mouth with many sharp teeth.

There were also marine crocodiles; to look in the face of a crocodile today is a fundamentally Jurassic experience, and it can't have been any funnier then than it is now. Amongst the invertebrates, several new groups appeared, including belemnites. These squid-like animals left behind bullet-shaped guards, about 15 cm (six inches) long, which fossilise easily.

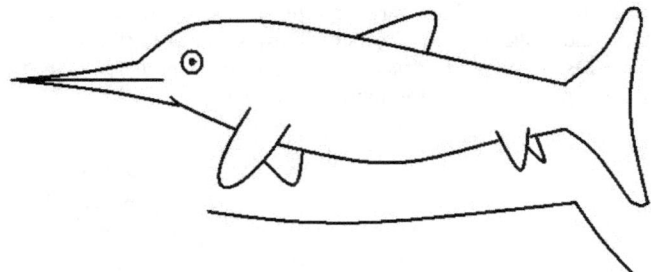

Ichy the ichthyosaur and below, the line of its spine

The ichthyosaur is one of the most fascinating of all Jurassic creatures. It was not fully understood until examples came from Holzmaden in south Germany in the later part of the nineteenth century. Many perfect specimens have now come from here, some of which prove beyond any doubt that these creatures gave birth to live young. However it is their uncanny similarity to fishes which has staggered the anatomists – the most perfect example known of what is called convergent evolution, where completely different types of animal assume the same form. Its bone structure tells us that the ichthyosaur is a reptile, but to any but a trained observer, it looks just like a swordfish. (Dolphins later accomplished a similar feat.) Only when the Holzmaden examples became available was it realised that the ichthyosaur had succeeded in evolving fins and a fishy tail, even though it has no bone structure to use for the fins, and only its spine for the tail. Hence many specimens have been found with a downward kink towards the end of the spine, which formed the bottom part of its arrow-shaped fishy tail.

Ammonites are the great zonal fossil of the Jurassic and indeed the whole of the Mesozoic. A detailed record of their morphologies and sequences was established by the middle of the twentieth century which enabled the era to be subdivided into time slices with a length of only 50,000 years.

The Jurassic was the golden age for the large herbivorous dinosaurs known as the sauropods – *Camarasaurus*, *Apatosaurus* (formerly called *Brontosaurus*), *Diplodocus*, *Brachiosaurus*, and many others. They grazed on the prairie flora of ferns and palm-like cycads, or the higher coniferous growth, according to their adaptations. The largest of them all was *Brachiosaurus*, heavily built with an enormously long neck and thought to weigh up to seventy tonnes. There is some evidence that there were even larger dinosaurs, perhaps weighing 150 tonnes! A mere nipper in this crowd was *Stegosaurus*, famous for the run of large plates down the back of its spine. The sauropods were preyed upon by large theropods, for example *Megalosaurus* and *Allosaurus* (*Tyrannosaurus* was Cretaceous). A theropod ("beast foot") is a bipedal dinosaur, mainly carnivorous.

The earliest dinosaurs – the saurischians ("lizard hipped") – still had hips like a newt, with the legs sticking out sideways. However a dinosaur innovation was the development of bird-like hips (ornithischian), from which their legs went vertically to the ground. All theropods belong to the earlier saurischian branch of the dinosaurs. The ornithischians dinosaurs were clearly much fast movers than the saurischians, and included heavily armoured herbivores such as *Triceratops* and *Stegosaurus*.

During the Late Jurassic, the first birds, like *Archaeopteryx*, evolved from small dinosaurs. The pterosaurs filled many ecological roles now taken by birds. Within the undergrowth were various types of early mammals, closely resembling modern small rodents. The oldest mammalian fossil yet found, a creature known as *Hadrocodium*, the skull of which was found in Yunnan, China, dates from about 190 million years ago – the early Jurassic.

In climatic terms, the aridity of the Permo-Triassic was replaced by a much wetter, lusher regime, allowing plenty of scope for the further evolution of the conifers, cycads, tree ferns and ginkgos which had survived from the Permo-Trias. (The maidenhair tree is the sole modern survivor of the ginkgos). In the Southern Hemisphere, podocarps (bushy evergreen conifers) were especially successful. These eventually evolved into the modern southern hemisphere flora of eucalypts and acacias.

****

Because most of the country was generally under 100 m/325 feet of water or so, Jurassic dinosaur fossils are a rarity in Britain, but in a way the climate is a relief from the Triassic as the country had drifted northwards and lay at about 30 – 40 degrees north. Jurassic rocks form a broad swathe of country from the Dorset coast, heading north-eastwards through Oxford and the Fens, to Lincolnshire and Yorkshire, where they terminate in a spectacular coast of high cliffs, for example at Ravenscar. There are also some outcrops in Scotland, on Skye and Mull, and in South Wales. The main English sequence is a series of clays, limestones and sandstones, all marine, and deposited in shallow seas according to distance from the Jurassic land masses to the north and west: inshore deltaic sands were laid down, then the finer clays, and finally the limestones deposited in tropical seas.

The main rocks are well-known: the Blue Lias, the inferior, middle and upper Oolitic Limestone, the Cornbrash, the Oxford Clay, the Corallian Limestone, the Kimmeridge Clay, and Portland (lime) Stone. The sequence gives the landscape an undulating appearance, the harder limestones forming some fairly substantial hills, with the softer clays forming vales between them. Although marked as clear sequences on the geological map of England, in any given locality there is likely to be a real jumble of shales, limestones, sandstones and clays. This type of sequence is known as a cyclothem and is typical of sediments laid down at the edge of the ocean, variably located over time.

The Blue Lias, which takes its name from the Dorset pronunciation for "layers", is made up of thinly interbedded shales and argillaceous (clayey) limestones. It is grey-blue in appearance and visibly cracked into layers. It is easily visible in outcrops on the Dorset coast and equally on the Yorkshire coast around Whitby. The formation is rich in ammonite fossils. One particular facies of it is known as "Shales with beef"! The thickest Lias strata in the country have been found in an unexpected place – directly off the coast of Cardigan Bay, where they abut the Cambrian Harlech Dome (the coast itself follows the fault line). The movement must have been thousands of feet on the downthrow side to leave the Lias below the Cambrian rocks. Another contemporary facies, occurring on the coast of north Yorkshire, is the Cleveland Ironstone (a type of sandstone), the basis of the steel industry still found there today.

The Oolitic limestone takes its name for the Greek for "egg" because it is made up of perfectly round spheres, each about a millimetre across, which look like eggs (or fish roe). Oolitic limestone from this era contains fossils but the main oolitic structure arises from inorganic grains rolling backwards and forwards on the floor of warm seas, coating themselves with carbonate, a process which continues today in tropical waters, and which is also found in Precambrian limestones. The stone outcrops from the Dorset coast between Weymouth and Lulworth, then north-eastwards past Cheltenham and Lincoln as far as the Humber. On the way it forms the Cotswolds overlooking the plains of Gloucestershire. The Oolites actually form three facies, Inferior, Middle and Great. It is one of the country's best known building stones, and whole villages in the Cotswolds – such as Broadway – are built of Inferior Oolite. Restoration work on the Houses of Parliament also replaced the original Magnesian limestone with limestone quarried from the Inferior Oolite in Rutland. The Great Oolite has been put to use in the construction of the wonderful cities of Bath and Oxford, and can be seen to advantage in the cathedrals at Wells and Lincoln. It can be cut to form flat ashlars which are used, for example, to form the creamy-white facings of Georgian buildings in Bath, and of Oxford colleges. The stone weathers to a honey colour, but the stones with a higher iron content are darker orange

In place of the Oolitic limestone in Yorkshire, there is an important sequence of estuarine or deltaic sandstones, resistant to weathering, which underlie the North York Moors National Park. These form a plateau of the bleakest of moors, but dissected by verdant dales, the main one being the valley of the Esk. The Estuarine Series here represents the deposits of a Rhine-like river which flowed from the land area to the north and east, known as the North Sea Dome.

In amongst the Oolites is the Cornbrash, an old English agricultural name applied in Wiltshire to a variety of loose rubble or brash which forms a good soil for growing corn (most of the other Cotswold rocks were only good for grazing). The name was adopted by William Smith for a band of shelly limestone. Although only a thin group of rocks (300 metres, 1025 feet), it is remarkably persistent, and may be traced from Weymouth to the Yorkshire coast. It crops out widely on the dip (eastern) slope of the Cotswolds because the dip of the rocks is parallel to the slope of the land.

Also found between Oolite beds is a layer of clay, sometimes hundreds of feet thick, known as Fuller's Earth, so-called because it was used to remove grease from wool in the fulling (felting) process in the woollen industry. It is in fact a bentonite volcanic ash containing the mineral montmorillonite which gives it detergent-like qualities.

Next in the sequence is the Oxford Clay, running north-east from Oxford through Peterborough and on to Lincoln in a broad band. The principal industrial use of this formation is brick manufacture – between a third and a half of all modern English bricks are made from it. Once again, it is rich in ammonites and other fossils, notably the oyster-like *Gryphaea*, a large clam known as the Devil's Toenail. This forms fossils so tough that they survive weathering and are washed away into other, newer rocks, where they can be misleading. They even appear as pebbles on the sea shore.

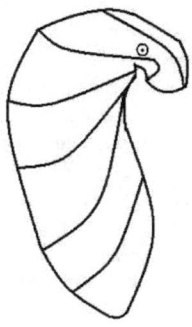

Griff the Gryphaea (eye added!)

Formations above the Oxford Clay include the Corallian Limestone, which, as its name implies, includes corals, and the Kimmeridge Clay, named after a village on the Dorset coast. Nearby is a famous feature, Lulworth Cove, where the sea has punched a hole through hard Jurassic strata to create a circular basin in the softer Cretaceous rocks behind. The rocks in this area have been subject to a surprising amount of buckling and folding.

The Kimmeridge Clay is distinctly bituminous and it is this rock which provides the oil and gas which is pumped up at Wytch Farm in Purbeck in Dorset. This is the largest onshore oilfield in western Europe. Although it only occupies a very small area, during the 1990s it produced 110,000 barrels of oil a day, and it is still productive. The Kimmeridge Clay also provides oil in the North Sea wells.

Folding at **Lulworth Cove**

There is another well-known landmark nearby, Chesil Bank (or Beach), a strand of large pebbles 27 km (seventeen miles) long, separated from the land by a lagoon known as the Fleet. At the end of the beach is Portland Bill, almost an island, and the source of the finest building material in the country – Portland Stone. This carries the largest of British ammonites, *Titanites*, which can reach 90 cm (three feet) across. The stone can be cut into slabs and is used to face the grandest of buildings, especially in London (as at St Paul's Cathedral and the British Museum), but also in many other places – for example it is used to face the Ashmolean Museum in Oxford. It is characteristically white in colour, though soon picking up pollution, as at St Paul's Cathedral. From neighbouring beds comes the Purbeck "marble" – in fact another facies of Jurassic limestone – which was used to build Salisbury Cathedral.

****

The Jurassic was the first great period for the fossil collectors. In 1825 a country doctor called Gideon Mantell published details of a Jurassic *Iguanodon* found by his wife poking out of the Sussex Weald. The year before, William Buckland had named a another Jurassic monster, *Megalosaurus*. Buckland (1784-1856) was one of the leading geologists of his day, and is also credited with the discovery of the Red Lady of Paviland (actually a man) in a cave in Wales, the oldest modern human bones found in Britain, dated to 33,000 years ago. Originally an old-fashioned Noachian Flood proponent, Buckland became a firm supporter of Louis Agassiz and his new theory of glaciation.

The best-known of the Jurassic fossil hunters was actually a woman, Mary Anning (1799–1847), who became famous for a number of important finds she made in the Jurassic marine fossil beds at Lyme Regis on the coast of Dorset where she lived. The world watched in amazement as she dug out previously undreamed-of, long extinct creatures, whose specimens fetched fancy princes from aristocratic collectors.

Mary worked on the Blue Lias cliffs. Her discoveries included ichthyosaurs, plesiosaurs (large marine reptiles with four flippers and a long neck), pterodactyls and some important fish fossils. When geologist Henry De la Beche painted *Duria Antiquior*, a famous scene of prehistoric life, he based it largely on fossils Anning had found, and then sold prints of it for her benefit. (He went on to become the first director of the British Geological Survey.)

Mary Anning was an uneducated seaside woman, and was never going to be able fully to participate in the scientific community of nineteenth-century England – dominated as it was by wealthy Anglican gentlemen. She struggled financially for much of her life. Her family was poor; her father, a cabinetmaker, died when she was only eleven. Nevertheless she remains well-known to this day.

****

The first real geologist of the Mesozoic era was William Smith (1769-1839), the man who produced the first geological map of England and Wales. He was born in the village of Churchill, Oxfordshire, the son of blacksmith. His father died when he was just eight years old, and he was then raised by his uncle. It is thought that he became interested in fossils at an early age, for the Jurassic strata of southern England abound in them. One had even found an everyday use. Known as a poundstone, it is shaped like a discus and is always 10 cm (four inches) across and weighs 620 grams (22 ounces). This matched the weight of a "long pound" used in the butter industry, so the fossils were used as weights. They are in fact echinoids (of genus Clypeus) – sea urchins – five-sided or round sea creatures, still found in the seas today. The name *Echinus* comes from the Greek and Latin word for hedgehog, as the living versions are covered in spikes.

In 1787, Smith found work as a surveyor's assistant at Stow-on-the-Wold, Gloucestershire. In 1791, he traveled to Somerset to make a valuation survey of the Sutton Court estate. He stayed there for the next eight years, latterly as an employee of the Somersetshire Coal Canal Company. (Note that there are no surface Coal Measures in Somerset – mine shafts had to be sunk to reach them.) As he observed the strata in the pits, Smith realised that they were arranged in a predictable pattern – the various strata could always be found in the same relative positions. So Smith began his search to determine if the relationships between the strata and their characteristics were consistent throughout the region.

Smith worked for the canal company until he fell into dispute with the owners and was dismissed in 1799. Thereafter he became a self-employed jobbing surveyor anywhere he could find work. In the course of his travels he would invariably take samples and fossils, and map the locations of the various strata, drawing cross-sections.

Smith was the first person truly to appreciate the significance of fossils. Up until his time, they had been collected – indeed, avidly collected – as interesting curiosities. Smith realised they were much more than this, as they could be used to identify sedimentary rocks definitively. Especially in the Jurassic sequences of southern England, there are repeated formations of sandstones, marls, clays and so on – often only a few feet thick – which look very similar. One sandstone could look just the same as another 150 metres (500 feet) higher in the sequence – but they would contain different fossils, and so could not be the same rock. Conversely, if a fossil assemblage found in Dorset matched another found in Yorkshire, it must be the same rock. Smith saw that the same succession of fossil groups from older to younger rocks could be found all over southern and eastern England.

In 1799 Smith produced the first large-scale geological map of the area around Bath, Somerset. After 1799 he was able to travel across the length and breadth of the country on surveying jobs, meeting some eminent people such as the agricultural reformers Thomas Coke of Holkham and the fifth Duke of Bedford along the way. In some ways, however, it was an unfortunate profession, as his wife was mentally unstable – described in fact as mad and bad – exhibiting nymphomaniac behaviour.

In 1815 Smith published the first full geological map of Britain. It covered the whole of England and Wales, and parts of southern Scotland. Symbols were used to mark canals, tunnels, tramways and roads, collieries, lead, copper and tin mines, together with salt and alum works. (Alum is a term for aluminium salts, used for purifying water and other cleansing.) The various geological formations were indicated by different colours; the maps were hand coloured. The map is remarkably similar to modern geological maps of England. A copy of it still hangs at the Geological Society in London, although protected from the light by a curtain. In the title of his biography of William Smith, Simon Winchester called this *The Map that Changed the World*. Some of the names Smith documented (like Cornbrash) are still used today, despite contemporary criticism.

Unfortunately for Smith, his maps were soon plagiarised and sold for prices lower than he asked for his own work. He had overstretched himself on his mortgage, and a quarry in which he invested lost money. In fact he proved much better at geology than he did at business. He went into debt and finally became bankrupt. In the summer of 1819 he was obliged to spend almost ten weeks in a debtor's prison in London. He returned to his London home of fourteen years to find a bailiff in occupation and his property seized. Smith then worked as an itinerant surveyor for many years until one of his employers, Sir John Johnstone, took steps to promote his name and work. Between 1824 and 1826 he lived and worked in Scarborough, and was responsible for the building of the Rotunda, a geological museum devoted to the Yorkshire coast. The Rotunda was re-opened as Rotunda – The William Smith Museum of Geology, in 2008.

In 1831 that the Geological Society of London conferred on Smith the first Wollaston Medal in recognition of his achievements. It was on this occasion that the President, none other than Adam Sedgwick, referred to Smith as "the father of English geology". (The Geological Society of London itself, at first little more than a gathering of aristocratic fossil-hunters, went from strength to strength as the marvels of the dinosaur age came to light, and was attended in its day by Palmerston and Gladstone.)

Although Smith never quite established himself in polite society, his nephew, John Phillips, whom he had brought up, did so. He eventually became Professor of Geology at Oxford, and introduced the concept of the fundamental division into Palaeozoic, Mesozoic and Cenozoic eras in 1841.

****

One famous episode of Jurassic palaeontology centres upon the Morrison Formation, a sequence of late Jurassic sedimentary rocks that is found in the western United States. This has proved to be the most fertile source of dinosaur fossils in North America. Most of the fossils occur in the siltstone and sandstone beds which form the relics of the rivers and floodplains of the Jurassic. It is centered in Wyoming and Colorado, with outcrops in many other western states and in Canada.

Thousands of dinosaur fossils have been recovered from this area, including the predatory theropod *Allosaurus*, and at least two species of *Stegosaurus*, first described by Othniel Marsh. Sauropods found here include *Diplodocus, Camarasaurus* (the most commonly found sauropod), *Brachiosaurus* and *Apatosaurus* (formerly known as *Brontosaurus*). The very diversity of the sauropods has raised some questions about how they could all co-exist. While their body shapes are very similar (long neck, long tail, huge elephant-like body), they are assumed to have had different feeding strategies, in order for all of them to have existed in the same time frame and similar environment.

When news of the discoveries in the Morrison Formation began to filter back east in the United States during the 1870s, it sparked of a period of intense rivalry known as the "Bone Wars". This involved two men, Edward Cope of the Academy of Natural Sciences in Philadelphia, and Charles Othniel Marsh of the Peabody Museum of Natural History at Yale. Marsh had a rich uncle, George Peabody, who had made his money in banking. He not only paid for Marsh's education, but for the Peabody Museum and the job of running it, which of course went to Marsh. Cope was also well-funded. All sorts of skullduggery was practised in the competition, including smashing up bones once a find had been cased.

The great aim of American palaeontology at this time was simply the naming of new species – for the name of the finder is generally attached to the species name and stays with it forever. There was much less interest in the hard grind of palaeontology, which involves identifying the fossils which can be used to correlate different beds in locations spread over a whole country or indeed the whole world. Dinosaur bones are not much use for that.

****

In the summer of 1966 I went to have a look at the Jurassic coast of Yorkshire for myself, on holiday with two friends, Martin and Stuart. I was only sixteen. Martin as a very tall boy, 1.9 metres (six foot four) at least, and good at cricket, but inclined to be fussy. Stuart was a thin, shortish boy with a pale face, deep brown eyes and short, dark hair. He looked as if he had a walk-on part in a children's play featuring mournful stoats and hedgehogs. He was fairly quiet, but he was also growing up, and had developed a pugnacious manner – rather like Bob Dylan, in fact, who was one of his heroes. He loved to tease me about my musical tastes.

"Gone out and bought the new Bachelors record then? Or was it the Cliff Richard?"

"Well I don't like blues."

"Why can't you listen to some real music? Like Dylan. Leadbelly. John Lee Hooker, Muddy Waters, Blind Lemon Jefferson, Howlin' Wolf."

"I don't see that stuff in the Top Ten."

"That is because the Top Ten is for idiots."

We set off to York on the day that England won the World Cup. We started to watch the game through the shop window of a TV rentals place, but the obliging manager let us inside the shop, and we saw the whole game. There were no customers anyway – everyone else was at home watching as well. We went on to Scarborough, Whitby and Robin Hood's Bay, staying in youth hostels. We had to fend for ourselves as far as most of the food was concerned, for the first time in our lives. At least we could manage scrambled eggs on toast, or thought we could. Stuart made his first, whisking up and egg, and putting this in a pan.

"Aren't you going to put some milk in with it?" I said.

"No – what for?"

"To make it go further."

When my turn came, I did add milk, and when it came out of the pan, it looked a lot better than Stuart's.

"John's got a lot more than you, Stuart," said Martin.

"Piss off," said Stuart.

After a few days, however, Martin became very fussy: "All this bread and stuff. We aren't eating properly – we're not getting enough protein!"

"You'd better get back to Mummy, quick!" said Stuart.

Even with only the O-level in Geology which I had recently passed, the Blue Lias in the cliffs at Whitby was blindingly obvious. It is this formation that contains inclusions of jet, which are polished up into ornaments and sold in the souvenir shops of the town. Jet is really a form of coal, derived from the remains of ancient monkey puzzle trees. Also at Robin Hood's Bay, I could discern that the cliffs were made of boulder clay. In fact the clay seemed to contain some amazingly large boulders of a completely different kind of rock. On revisiting the site years later, I came to the conclusion that it was indeed boulder clay, but that these same boulders had been dumped at the base of the cliff by the local authority, to try to preserve it from erosion! Martin and Stuart were not in the least bit interested in my geologizing. At the youth hostel at Boggle Hole, near Robins Hood's Bay, however, I met someone who was – a professional geologist, from Cambridge! He knew all about the Millstone Grit cliffs overlooking the Calder Valley in my home town of Halifax as well. They are certainly a magnet for climbers but I can't say I've ever noticed a geologist there.

The fact that I came to pass O-level Geology was in itself somewhat fortuitous. I certainly learnt plenty of geology for the examination, and was indeed the only boy in my school to pass it with a grade one. I seem to remember spending rather a lot of time learning about the internal architecture of lamellibranchs and brachiopods. The problem had been one of staying in the class for the whole year. Geology was taken as an optional extra by boys in the Lower Sixth form (sixteen and seventeen year-olds). Those eligible for the class were all the scientists and the geographers. Geology had only recently been introduced, as a new physics teacher, a Mr Greenwood, a fresh face to replace the dinosaur who had been at the school since my father's day, had appeared on the scene. He was really a geologist rather than a physicist, and volunteered to take the O-level class, which took one year as a minor subject. He was an irascible man, about thirty years old. He was faced with a class which he thought was far too large – there must have been thirty of us, out for an easy O-level – most of whom showed a marked disposition to lark about in class. So Greenwood soon set about a weeding process.

I sat at the back of the class with Martin and larked about as much as anybody, and frequently caught the exasperated glance of Mr Greenwood, but there seemed to be a problem. He didn't actually know the names of anyone on the back row. He only knew the names of the people doing Physics A-level, who sat nearer the front. Pretty soon it was – "You – Bates – out!" Next week "Rastrick – out!" Bates,

Rastrick and the rest trooped gloomily out of the class, and were not allowed back. Martin and I survived – I think he may have marched round once to actually demand our names, but got no further. I don't recall that more than one or two physicists sat the exam, and they looked utterly bewildered!

A good place to see geology in action – crater lakes on the island of **San Miguel** in the volcanic Azores

Chapter 14 – Cretaceous

The Cretaceous, a name derived from the Latin "creta" (chalk) is the geological period running from about 145 to 65.5 million years ago. It is the third and last period of the Mesozoic era, and at 80 million years, the longest period of the Phanerozoic Eon. It was first defined as a separate period by a Belgian geologist called Jean d'Halloy in 1822, in reference to the chalk strata of the Paris Basin.

The chalk is the world's most widespread single sedimentary rock. The high eustatic (worldwide) sea level and generally warm climate of the Cretaceous meant a large area of the continental shelf was covered by warm, shallow seas. These proved to be ideal conditions for chalk deposition. Chalk itself is calcium carbonate deposited by the shells of marine invertebrates. The larger of these are foraminiferans, typically a millimetre across. They are embedded in a dust which under the microscope can be seen to be made up of coccoliths, the remains of marine algae – minuscule plants – that flourished in the Cretaceous seas. Within the chalk are typically found bands of flint. Not every marine creature made a shell of calcium carbonate – others, especially radiolarians, used silica (silicon dioxide) and this equally organic material came to be concentrated in distinct bands of flint, often collected around the remains of sponges and other marine creatures. The Chalk does contain larger fossils, typically sea urchins (echinoids).

However, there is one mystery about the Chalk, and one which seems to defy Lyell's principle of uniformitarianism – although it was a widespread deposit on the continental shelves during the Cretaceous, it has proved difficult to find an area of the world where it is forming today. The only place which accumulates a creamy mud of calcareous shells is the bottom of the ocean, and that will never become dry land. Also, it only occurs once in the stratigraphical column (unlike, say, limestone). There must have been something special about the environment when the Chalk was laid down.

During the Cretaceous, Pangaea completed its tectonic disintegration into the separate continents we would recognise today. The Iapetus Ocean of the Silurian was destined to reappear when Laurentia began to separate from Europe in the Cretaceous, more or less – but not quite – on the line of the original suture. Its new name would be the Atlantic. The southern part of the North Atlantic began to open out in the

Cretaceous, but in the northern part this was a Tertiary process. In North America, tectonic movements forced up the Rocky Mountains and high plateaus of the west. In fact the Rockies have been built not by one but by a series of orogenies over as much as 200 million years. The latest phase, the Laramide orogeny, began in the late Cretaceous, about 75 million years ago.

In the southern hemisphere, Gondwana split up into separate continents and the South Atlantic and Indian Oceans formed. Henceforth each part of Gondwana began to evolve is own distinctive flora and fauna. This is ultimately the reason why Madagascar is full of lemurs, Australia of marsupials and eucalypts, and why South America was to develop such a cornucopia of distinctive plants and animals.

Of all the continents, Australia (plus New Zealand) has ended up with the strangest fauna. Nowhere else is there anything like a kangaroo, described by one early English sailor as an "eighty-pound mouse". Only here are monotremes found (the duck-billed platypus and the echidna), as well as koalas and wombats. Australia does have some placental mammals including mice, bats and rats, but the dingo is a recent introduction from New Guinea. Still it all has an old-fashioned look, brought about by 120 million years of separation. In fact Australia spent much of the Mesozoic in sub-polar latitudes, along with Antarctica, where the winters were cold and no sun shone for two months of the year. Still it had a fauna of dinosaurs, even quite large ones, though no one is quite sure how they survived the winters. South America had a much more varied fauna, but again, retains its strangeness to this day. For example its monkeys are quite different from Old World varieties, and it has the sloth and the armadillo.

Another result of the splitting up of the southern part of Pangaea was the creation of a "large igneous province", similar to the Siberian Traps, which surfaces at Kerguelen Island in the Antarctic. Massive outpouring of basalt occurred here over a long period, from 139 to 34 million years ago.

The active drifting of the Cretaceous lifted great undersea mountain chains along the welts, raising eustatic sea levels worldwide. Broad shallow seas advanced across the lower parts of central North America and Europe, leaving thick marine deposits behind. At the peak of the Cretaceous transgression, one-third of Earth's present land area was submerged. North America itself formed three large islands divided by the Western or Cretaceous Interior Seaway, running all the way from the Gulf of Mexico to the Arctic, and the Hudson Seaway. Sea levels were 200 metres (650 feet) above modern levels (from Tim Haines,

Walking with Dinosaurs – note that a quarter of modern North America was flooded.)

Marine transgressions and their opposite, regressions, punctuate geological history. Given that, in the absence of an ice age, the amount of water in the world's oceans must remain more or less constant, they can only realistically come about through changes in the relationship between the ocean floor and the continental crust. In other words there must be a change in the capacity of the sea-basin. It is known that submarine mountain-building is a variable feature of geological history, which as noted may have been especially active in the Cretaceous. The material to build those undersea mountains must come from somewhere. If it is drawn from adjacent areas of mantle in the middle of the oceans then it is hard to see how this would ultimately affect sea level. If however the material is drawn from underneath the continental crust, then the continents will subside, and this will cause sea levels to rise.

Another mechanism which could bring about a marine transgression is a falling-off of the rate of subduction, which creates massive "holes" – deeps such as the modern Challenger Deep – in the ocean. Subduction occurs where an oceanic plate dives under a continental one, but does not happen on every continental border. The alignment of the continents relative to the oceans at the start of the Cretaceous could have meant that little subduction was taking place.

The average fluctuation of sea level in a major transgression or regression is, in terms of the radius of the earth, tiny – hundreds of metres at the most, when the radius is 6371 kilometres (3959 miles), a matter of one thousandth of a percent. It is possible that such small changes could be brought about by changes in the internal magnetism of the earth, "sucking in" one part of the crust (and so causing a transgression of the sea), or alternatively pushing it out. Variations in the earth's magnetism are certainly known, such as the current South Atlantic Magnetic Anomaly, a weakening of the earth's magnetism in that area. I personally have not seen this phenomenon discussed with respect to sea levels, but, in the case of localised (as opposed to eustatic or worldwide) ones it would seem to be a possibility.

However, the most obvious cause of a marine transgression is the reduction of the land area to peneplain (almost flat) status. It is known that the erosion of the modern Himalayas is actually filling up the Bay of Bengal and Arabian Sea either side of the Deccan. By the time of the Cretaceous, it had been a very long time since the last major orogeny, the Hercynian – in fact over 150 million years. The continents were busy splitting up, not bumping into and undermining one another. The mountains on the land had made their way into the sea.

Flowering plants (angiosperms) spread during the Cretaceous, to become the dominant class of plants by the end of it. Their advance was helped by their use of co-evolving insects – bees, moths and butterflies – for pollination purposes. Gymnosperms from the earlier Mesozoic periods also continued to thrive, including the monkey puzzle tree.

On land, marsupial mammals evolved early in the Cretaceous, and true placentals later in the period. However the fauna was of course dominated by dinosaurs, which were at their most diverse stage. Sadly the wonderful large sauropods of the Jurassic had disappeared, except in South America, where they continued to evolve. Towards the end of the Cretaceous, duck-billed dinosaurs known as hardrosaurs were by far the most common types, but there were also many horned dinosaurs including the famous *Triceratops*, which, as its name implies, had three horns, and looked like a mean critter! It would need to be, with *Tyrannosaurus rex* for company. *Tyrannosaurus* is the ultimate theropod – even its gigantic bird-like footprint is enough to terrify. The Hell Creek formation of Montana includes many *Triceratops* remains. (Incidentally the 1995 film *Jurassic Park* might better have been called *Cretaceous* Park, as most of the dinosaurs in it were indeed Cretaceous in age.)

The flying pterosaurs were common in the early and middle Cretaceous, filling many ecological niches since occupied by birds. Some were small, thrush–sized creatures, others prone to gigantism. The largest of them all, *Quetzalcoatlus*, named after a fearsome Aztec god, had a wingspan of fifteen metres (50 feet)! There has been much speculation about how it managed to get off the ground. However as the Cretaceous proceeded the pterosaurs faced growing competition from birds, and their number thinned out markedly. The Jehol *lagerstätte* in Liaoning, NE China, provides a glimpse of life in the early Cretaceous, where preserved soft tissues of numerous types of small dinosaurs, proto-birds, and even mammals have been found. These include the feathered creatures *Microraptor* and *Confuciusornis*, intermediate between dinosaurs and birds.

Cut in Cretaceous sandstone, the **gorges of the Elbe** above Dresden make great scenery, and also good climbing.

****

It is thought that given the depth of the Cretaceous inundation at the base of the chalk, put at 200-300 m/650-975 feet, chalk was deposited over much of the British Isles and indeed north-western Europe. For example it is thought to have covered West Yorkshire, where there is no sign of it now, and there is block of it the size of a house buried by Tertiary volcanic rocks on the Scottish isle of Arran, where again there is otherwise no trace of the rock. Today it occupies a broad swath of country in the south and east of England. It offers an easily-recognisable landscape of rounded hills and dry valleys, thought to have formed when periglacial conditions (that is, at the margins of the ice sheets) made the rock impermeable. The rock itself weathers easily as it lacks the mechanical strength of say Portland Stone, it is porous and so subject to chemical weathering, and it is also heavily jointed. This feature exposed it to the freeze-thaw action of ice, a very powerful agent of erosion. Ice occupies 9 per cent more volume than free water – we are all familiar with its effects at home! In fact the characteristic landscape of the Chalk may well have been created in a periglacial environment, and may not be changing much now.

The chalk itself forms the hills of the North and South Downs. These represent the edges of a denuded anticline, the Weald. Inside the Downs are older rocks, laid down before the chalk, but now exposed because of the erosion of the dome above them. The Downs start on the coast at the White Cliffs of Dover, then lie westwards along the coast past Beachy Head, swinging inland in West Sussex and round, south of London and back to Dover. The chalk branches off to form Salisbury Plain and the Marlborough Downs to the north-west, and the Chiltern Hills to the north of London. The outcrop continues into East Anglia, reaching the coast at Hunstanton, where there is an unusual formation of red chalk, stained by iron. The rock is not finished yet, however, as it reappears in eastern Lincolnshire, and then in Yorkshire as the Yorkshire Wolds. Here it strikes the coast in high cliffs at Flamborough Head, where mighty waves crash into the formation, creating sea stacks and caves. It is one of England's great rocks, used by mankind from time immemorial as it offers a light and easily cultivable – if thin – soil.

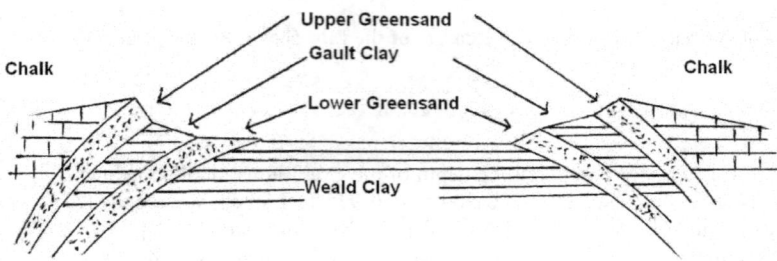

Schematic representation of the Weald, a denuded anticline. The Chalk would once have formed a cap over all the older rocks beneath it, but now forms the high ground – the North and South Downs – overlooking the hollowed-out centre of the anticline.

Inside the Weald the oldest formation is a series of non-marine rocks, probably laid down in a delta, and known as the Weald Clay. Some of these rocks contain iron nodules, and so formed the basis of an iron-smelting industry which continued for centuries in the Weald. This facies also contains the remains of *Iguanodon*, and a very complete specimen, "Claws" (officially *Baryonyx* – though from the shape of its teeth a vegetarian, it was obviously prepared to defend itself! – was discovered in 1983, chipped out and put on display in the Natural History Museum in London. There follow three marine facies, the Lower Greensand, the Gault Clay and the Upper Greensand, before the Chalk itself. The Lower Greensand is not normally green, but yellowish, and forms poorish hilly country. Its original green colour derives from the presence of the mineral glauconite, which oxidizes to yellows and reds when exposed to weathering. The Upper Greensand actually does have a greenish tinge, and weathers out to produce a rich, loamy soil long sought out by farmers. The Gault Clay is quite a widespread formation in southern and eastern England where it lies across much older rocks in unconformities, forming vales beneath harder lithologies. As it is an impermeable rock, and the Chalk above it is permeable, the junction between the two throws out streams and creates rockfalls over the slippery Gault Clay.

The rocks of the Weald surface again in the southern part of the Isle of Wight. The Chalk is folded vertically to form the central hills of the island. North of it are younger rocks, and south of it the Wealden facies. The Chalk itself runs out to sea in the famous formation of the Needles, three sea stacks on the western part of the island at Alum Bay.

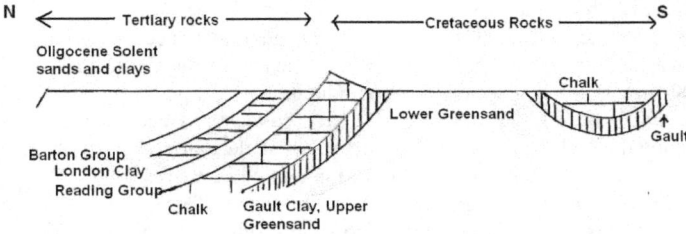

Cross-section of the geology of the Isle of Wight. The Chalk forms a ridge running E-W across the centre of the island. The younger rocks lie in the north, but the succession is rapid as the island is only 21 km (13 miles) from the northern to the southern tip.

****

There was a decline in biodiversity during the later stages of the Cretaceous, which nevertheless appears to have ended with an almighty bang – the impact of a meteor, nine kilometres (5.6 miles) across. Few species of fossils pass through the K–T boundary (Kreide (chalk)-Tertiary). In those places where there is continuous deposition of sedimentary rock through the K-T boundary, the transition appears very sudden.

Evidence for this was first found at Gubbio in Italy by the American geologist Walter Alvarez. He found extraordinarily high levels of the element iridium – ten times the expected amount – in a thin layer of clay deposited at the K-T boundary. Iridium is comparatively rare on earth, but is common in meteorites. In 1980 Alvarez published a paper in the American journal *Science*, jointly with his father, Luiz Alvarez, a Nobel prize-winning astronomer (this helped to ensure publication!) Professional geologists were of course sceptical at first, but since publication, the evidence has mounted to the point where it has become overwhelming. Iridium layers at the K-T boundary have been found worldwide – in Tunisia, France, Denmark and Texas for example. From some of these sections, "shocked quartz" has been recovered, indicating its deformation at high pressure. The regular flora apparently disappeared utterly, to be replaced by ferns – the "fern spike" – which are well-known as a pioneer plants because they spread spores so easily. Then in 1991 the smoking gun was finally identified – a crater of an appropriate age, at Chicxulub on the Yucatan peninsula in Mexico, 200 kilometres (125 miles) across.

Despite the accumulating evidence, not all geologists accept that a meteor was responsible for the catastrophe at the end of the Cretaceous, partly because there is another strong candidate. At the same time, in the area that is now India, massive lava beds called the Deccan Traps were erupted. The Traps are on another scale altogether from Hawaiian-style shield volcanoes. 2.5 million cubic kilometres (600,000 cubic miles) of basalt were erupted over a time span of probably only 1.5 million years. This is enough to cover the whole of the area of Alaska and Texas to a depth of one kilometre. In addition, this material was poured out on top of existing continental rock, not onto the ocean floor. This also happened at the eruption of the Siberian Traps, which brought down the curtain on another great extinction at the end of the Permian. These continental traps deny the principles of Lyellism, because nowhere on earth today can this phenomenon be observed. They share their non-uniformitarianism with the Chicxulub meteor, but there can be absolutely no doubting the evidence. It is certainly striking that the two greatest mass extinctions – out of five altogether – have coincided with these truly vast lava eruptions. Also, eruptions of this type bring up unusually large quantities of iridium from the mantle.

The cause of the eruptions is debated. It is thought that the Indian subcontinent passed over a hot spot – a mantle plume – which burst through the crust. The island of Reunion is now sitting over this spot. Yet there are more recent, if smaller, continental basaltic eruptions which do not appear to have brought life to a juddering halt. One such forms the Columbia River Plateau in the north-west of the United States, covering 130,000 square kilometres (50,000 square miles) and dating from only seventeen million years ago. Another similar place is the Snake River basin of southern Idaho, where basalts cover 50,000 square kilometres (20,000 square miles) and are of Quaternary age.

There is a third explanation for the disaster at the K-T boundary – a worldwide fall in sea levels, enough to increase the land area of the continents by a quarter. One can't help feeling that this happened because the lava which had been beneath the crust in the Deccan was now on top of it, leaving a large hole to fill in the mantle! This would cause the shallow seas to withdraw from the continental shelf. Also – can it be a coincidence? – exactly this set of circumstances was also found in the greatest extinction of them all, at the end of the Permian, when the Siberian Traps were created and there was also a marine regression.

The effects of blocking out the sun for a period of months or years are not difficult to imagine, but the lack of sunlight was evidently not the only problem. The mean world temperature would have plummeted rapidly, itself accounting for many fatalities. Some writers have postulated torrential rains of sulphuric acid. So the phytoplankton in the sea perished, destroying the base of the food chain; the same thing happened to the green plants on the land. Herbivorous animals, which depended on plants and plankton as their food, died out as their food sources became scarce; in turn, top predators such as *Tyrannosaurus rex* also perished. Foraminifera and molluscs, including ammonites, as well as organisms whose food chain included these shell builders, became extinct or suffered heavy losses. Nothing weighing more than 35 kg (80 lbs) survived. Most famously, the dinosaurs disappeared forever.

Despite the severity of this boundary event, there was significant variability in the rate of extinction between and within different phyla. Omnivores, insectivores and carrion-eaters survived the extinction event, perhaps because of the increased availability of their food sources. At the end of the Cretaceous there seem to have been no purely herbivorous or carnivorous mammals left. Mammals and birds which survived the extinction fed on insects, larvae, worms, and snails, which in turn fed on dead plant and animal matter.

In the oceans, extinction was more severe among animals living in the water column than it was among animals living on or in the sea floor. Animals in the water column are almost entirely dependent on primary production from living phytoplankton, while animals living on or in the ocean floor feed on detritus or can switch to detritus feeding.

The largest air-breathing survivors of the event, crocodiles and similar creatures, were semi-aquatic and able to scavenge. Modern crocodilians can live in this way and can survive for months without food.

Needless to say, the level of biodiversity took a long time to recover after such a disaster.

Chapter 15 – Paleogene

The Phanerozoic used to be divided into four great eras, the Palaeozoic, Mesozoic, Tertiary and Quaternary, of which the Quaternary was much the shortest at only 2.58 million years. However recently the idea of the Tertiary, or third era, has become unfashionable, though the term is still widely used. Current practice is to divide it into the Paleogene (65 million years ago to 23 million years ago) and the Neogene (23 million years ago to 2.58 million years ago). The old Tertiary divisions, the Paleocene, Eocene, Oligocene (Paleogene), Miocene and Pliocene (Neogene) used to have the full status of periods, but have now been downgraded to smaller divisions of time called epochs. Following the Neogene there is still the Quaternary, divided into the Pleistocene and Holocene or Recent epochs as before. The whole span of time from the end of the Cretaceous to the present day is known as the Cenozoic era.

The Paleocene, the "early recent", is the geological epoch that lasted from about 65.5 to 56 million years ago. It immediately follows the mass extinction event at the end of the Cretaceous, and its name describes the early and (not surprisingly) somewhat lacking fauna that arose during the epoch, before modern mammalian orders emerged in the next epoch, the Eocene. It is as if the great actors have gone from the stage, and left only midgets behind.

The early Paleocene was cooler and dryer than the Cretaceous, though it was still warm by today's standards, with ice-free poles. Temperatures then rose sharply as the epoch wore on. The climate became warm and humid worldwide towards the boundary of the next period, the Eocene, with subtropical vegetation growing in Greenland and Patagonia, and crocodiles swimming off the coast of Greenland. In the subtropics, climates were hot and arid, as they are today.

The big northern continent had not yet fully separated. Europe and Greenland were still connected, and North America and Asia were intermittently joined by a land bridge, but Greenland and North America were beginning to split apart. South and North America remained separated by tropical oceans. Africa headed north towards Europe, slowly closing the Tethys Ocean which lay between them, and India began its migration towards Asia. This would lead to a tectonic collision and the formation of mighty mountains, the Himalayas, in the Miocene. The inland seas in North America (Western Interior Seaway) and Europe had receded by the beginning of the Paleocene, making way for new land-based biota.

In the seas, the marine reptiles had of course gone, and there were as yet no marine mammals such as whales. Sharks became the top marine predators.

The Paleocene was a unique time, a cornucopia of ecological opportunities provided for any branch of the terrestrial fauna to take advantage of the enormous gaps left by the demise of the dinosaurs. The groups which leapt into this gap were the mammals and the birds. A host of brand new designs was created to meet the opportunity. This did not happen amongst other survivors from the catastrophe – turtles, lizards and crocodiles, for example. Tiny shrew-sized mammals grew to the size of dogs within three million years. Unfortunately for the palaeontologist, these creatures do not fossilize well, but there is one important thing they left behind – their teeth. The Paleocene palaeontologist has to know a lot about teeth, for there is little else left of these early mammals.

While early mammals were small nocturnal animals that mostly ate soft plant material and small creatures such as insects, the demise of the dinosaurs allowed them the ecological space to grow bigger and occupying a wider variety of niches. Ten million years after the death of the dinosaurs, there were many species of medium sized mammals scavenging in forests, and large herbivorous mammals and even carnivorous mammal hunters. However the mammals were still far from modern forms that we would recognize.

Birds diversified in the Palaeocene. One notable category is made up of large, carnivorous flightless birds (also called Terror Birds), including the fearsome *Gastornis* in Europe. Reptiles including snakes and crocodiles grew to enormous sizes, perhaps encouraged by the warmth. *Titanoboa*, found in a coal mine in Mexico, is the largest snake ever known at an estimated 15 metres (48 feet)!

Geology is of interest to oil companies above all others, and they in turn are likely to be most interested in rocks of Mesozoic or Cenozoic age. An underground anticline with a cap of impermeable rock and underlying permeable rock can form a trap for oil and natural gas. It was the requirement to map and correlate these underground beds which led to a breakthrough in Tertiary palaeontology. The great zonal fossils of the Cenozoic are as unlikely as the graptolites of the Palaeozoic, for they are foraminiferans, single-celled floating creatures, which are used for this purpose. The first people to discover their usefulness were the Russians, who had discovered new deposits of oil in the Caucasus immediately after the Second World War. There was an urgent need to correlate the different beds, and foraminiferans served the purpose.

Exactly the same need was identified by western oil companies at Sirte in Libya, where oil is found in Cenozoic deposits, but it was found that the Russians had a head start. In fact their palaeontologists were all women, and some of the forams are named after two of these women, Valentina Morozova (*Mozorovella*) and Nina Subbotina (*Subbotina*).

****

Tertiary rocks even from the Paleocene are generally soft and, with some exceptions, can usually be scraped out with a trowel. The most notable exceptions are sarsen stones, sandstones which have become indurated by a silica cement. These were used to make the big uprights and lintels of Stonehenge, and also the standing stones of the stone circle at Avebury, and they date from the Paleocene. However the very softness of Tertiary rocks has created some wonders of geomorphology around the world, as they weather easily and can leave fantastic shapes if a hard rock overlies a softer one.

There is no better place to see this effect than in Bryce Canyon National Park in southwestern Utah in the United States. The rim at Bryce varies from 2,400 to 2,700 metres (8,000 to 9,000 feet). It is distinctive due to geological structures called hoodoos, formed by wind, water, and ice erosion of weak sedimentary rocks. The red, orange, and white colors of the rocks provide spectacular views.

Bryce Canyon, Utah

The Bryce Canyon area was settled by Mormon pioneers in the 1850s and was named after Ebenezer Bryce, who built his log cabin in the area in 1874. Bryce Canyon was not formed from erosion initiated from a central stream, meaning it technically is not a canyon. Instead headward erosion by streams has excavated large amphitheater-shaped features in the Tertiary-aged rocks. This erosion exposed delicate and colorful pinnacles, the hoodoos, that are up to 61 metres (200 feet) high. A hoodoo (also called a fairy chimney, a much more descriptive name!) is a tall, thin spire of rock that protrudes from the bottom of an arid drainage basin or badland. Hoodoos consist of relatively soft rock topped by harder, less easily eroded stone that protects each column from the elements. They typically form within sedimentary rock and volcanic rock formations. Hoodoos are also a tourist attraction at Goreme in the Cappadocia region of central Turkey where whole houses have been carved within them. Similar structures can be created by wind erosion in deserts, known as pedestal rocks.

As I personally like to include a geological component to my holidays, rather than sitting on a beach, I have been to both Bryce Canyon and Cappadocia. Apparently Bryce Canyon doesn't get many visitors, and it is certainly remote, but one reason could be that there is no need to get down amongst the hoodoos to appreciate the fantastic scenery. There are viewpoints above the natural amphitheatre from which the whole can be viewed. It is one of the many geological marvels of Utah, but there are drawbacks to the state as well. I seem to recall it was rather difficult to buy any wine in the restaurants, and when finally located, it was 6% alcohol! What a place! That's the Mormons for you, but it's their state. I once had a worse experience in Finland, where the sale of alcohol is also restricted. It is impossible to make any sense of Finnish, so you can't read what is says on the can, but I will admit to being a bit suspicious when I found I was able to buy six small cans of beer in a supermarket. Only after spending most of an evening drinking this stuff did I realise that I had been drinking alcohol-free beer.

We approached the Cappadocia area of Turkey by heading north-eastwards from the coast at Dalaman. I have seen the south and west of Turkey – Istanbul and the coast – as many people have, and know it to be relatively green, steamy even in summer. Flying across the northern and eastern parts of the country, the view is very different, of snow-capped mountains, the source of mighty rivers, the Tigris and the Euphrates, and Mount Ararat, the biblical home of Abraham. Yet crossing inland through Konya to Goreme the land is completely flat, and also semi-arid: the Anatolian Plateau. In fact I have never seen such an expansive area of completely flat land anywhere in my life. I speculated on the reason for it at the time. I do not know the answer, but this is an area of inland drainage, and there are two large salt lakes in the region today. I suspect that at one time, and in a much wetter climate, these lakes would have covered much greater areas, and the flat land of Anatolia is a former lake bottom. If so, those lakes must have persisted for very long periods.

Fairy chimneys at **Goreme, Cappadocia**, Turkey
****

The Eocene epoch (56-35 million years ago) spans the time from the end of the Paleocene to the beginning of the Oligocene. The end is set at a major extinction event which may have been caused by the impact of one or more meteorites in Siberia and in Chesapeake Bay in the United States. The name *Eocene* comes from the Greek, meaning "new dawn" of mammalian creatures that appeared at this time.

The Eocene includes the warmest climate in the Cenozoic Era, starting at the Paleocene–Eocene Thermal Maximum. This is a sudden spike in temperatures, thought to have gone up by a worldwide average of 6-8 degrees Celsius in just 20,000 years, to a sky-high 28 degrees. The phase lasted about 200,000 years altogether; then temperatures fell back to their previous levels, but climbed once again, back to 28 degrees in the middle of the "Eocene Optimum" spanning the central millions of years of the epoch. The temperature then gradually fell to 16 degrees at the end of the Eocene, to 12 degrees during the Ice Age, and 14 degrees today. At the Paleocene-Eocene Thermal Maximum, the oceans turned acidic as a result of absorbing large amounts of carbon dioxide.

There has been much speculation into the causes of the Paleocene-Eocene Thermal Maximum and research on the subject continues to today. The earth is known to contain large quantities of sequestered methane hydrates, frozen in Arctic soils or beneath areas of high pressure in the oceans. They are both widespread and unstable. An increase in ocean bottom temperatures of 5 degrees Celsius or a decrease in overlying pressure of a couple of pounds per square inch would be enough to release large quantities of methane into the atmosphere. Whilst the first of these seems unlikely, given that ocean bottom temperatures hover not much above zero, the second is possible, as it would represent only a small fraction of the weight of the ocean column above the methane hydrates. So some researchers have invoked them as the source of the terrific heat wave at the end of the Paleocene, as methane is a very effective greenhouse gas. Cores retrieved from across this boundary do show sudden blips in the ratios of light to heavy carbon and oxygen isotopes. In the case of carbon, this favours the lighter isotope which is predominant in methane produced by the natural processes of decay. One proposal is that the methane hydrates were disturbed by submarine earthquakes, though in fact, these are commonplace.

If millions of tonnes of methane were released into the atmosphere over 20,000 years, then it clearly took only a short period, in geological terms – 180,000 years – to put the genie back in the bottle, before the whole process repeated itself over a much longer timescale.

Australia and Antarctica started to split apart in the Cretaceous, about 99 million years ago. By the Eocene epoch 45 million years ago, Antarctica was heading inexorably for the south pole and cooling down. To be on the wrong side of this divide was bad news for any plant or animal, and there must have been a lot of them. In the northern hemisphere, Laurasia continued to break up, as Europe, Greenland and North America drifted apart and the Atlantic opened out.

At the beginning of the Eocene, the high temperatures and warm oceans created a moist, balmy environment in the most unlikely places. Even polar forests were quite extensive. Swamp cypresses were to be found on Ellesmere Island and magnolias grew in Alaska. Newly-evolved, broad-leaved flowering trees of the type which fill the temperate and tropical forest today became common. There has probably never been a time when the earth looked so verdant.

Cooling began mid-period, and by the end of the Eocene the continental interiors had begun to desiccate. Deciduous trees filled the temperate forests of the world.

The oldest known fossils of most of the modern mammal orders appear during the early Eocene. Dwarf forms were dominant, generally under 10 kg (22 lbs). Based on comparisons of tooth sizes, Eocene mammals were only sixty percent of the size of the primitive Paleocene versions that preceded them. It is assumed that the hot Eocene temperatures favored smaller animals that were better able to manage the heat. Early forms of ungulates (hoofed animals), primates, rodents and marsupials appeared. The sudden appearance of bats, one of the first distinctive modern mammal groups to evolve, leaves their origin a mystery. There were many species of elephants, none larger than a small pig. Amongst the mammals there was a distinct difference between the placentals of the northern hemisphere and the marsupials of the southern. Several rich fossil insect faunas are known from the Eocene, notably preserved in Baltic amber.

In fact it was after a rather slow start in the Paleocene that mammal radiation – the proliferation of new species – picked up momentum in the Eocene. The result was eventually to be a dazzling array of new creatures, not just plain old horses, cows, gazelles and pachyderms (elephants, hippos, rhinoceros), but truly odd mammals – kangaroos, camels, tree sloths and giant South American rodents. One of the commonest of the new large grazers was the brontothere, the size of an elephant but horned like a rhino, the bones of which have turned up in Mongolia and also in large numbers in Wyoming. Another creature from this epoch is the entelodont, an ugly-looking brute with vicious teeth, resembling a warthog but weighing nearly a ton in larger specimens. Its skulls are often found pierced with tooth marks, thought to have been inflicted by other entelodonts of this clearly aggressive breed. Even more fearsome is *Andrewsarchus*, only one skull of which has ever been found, in Mongolia – but what a skull. With massive teeth, it must have been one of the largest predators of the Cenozoic, thought to weigh three tonnes. It is named after the man who found it, Roy Chapman Andrews, an American fossil-hunter who was reputedly the inspiration for the fictional Indiana Jones. He set out for the Gobi desert with a large party and almost immediately found outstanding fossil beds at Hsanda Gol.

Also in the Eocene roll call were anteaters with long snouts and sticky tongues which evolved into more really odd mammals, pangolins and aardvarks. Pangolins have barely changed in 50 million years and looked just as peculiar as a modern example with their strange armour of interlocking, triangular plates. Another creature from this time is *Ambulocetus* ("walking whale"), one of those elusive missing links. It is thought to be the true ancestor of the whales, but retained the use of its legs, living in and around terrestrial waters. Several specimens have been dug up in Pakistan. It moved in water by flexing its spine up and down, in the manner of an otter, instead of side to side, like a fish – a usage which was inherited by all whales.

In the Eocene oceans, early marine mammals appeared, including *Basilosaurus*, at 18 metres (60 feet) long, the largest known Eocene mammal. Despite its name ("king reptile") this was not a reptile but a primitive whale that is thought to be descended from land animals that existed earlier in the Eocene. It still retained the relic of its hind limbs. Remains of it surfaced in Louisiana before more were found in the famous fossil beds at Fayum in Egypt, now in the middle of the Sahara Desert. Also in the sea appeared fantastic dugongs and manatees, seals and porpoises.

The diversity of the Eocene fauna is recorded in marvellous detail in one of the world's best *lagerstätten,* the Messel shale from near Frankfurt in Germany. This was the site of an Eocene lake 50 million years ago, where the remains of the fauna were preserved in soft mud. Every bone of even delicate creatures such as bats has been preserved; seven species have been identified, indicating an already long evolutionary history. There are the fine remains of frogs, toads, turtles, six different kinds of crocodile, millipedes and constricting boas. (Note that venomous snakes did not evolve until the Miocene, and that the ability to produce venom was never repeated in any mammal species.) Also preserved is *Formicium giganteum*, the largest ant ever found. The Messel has birds – flamingos, owls, nightjars and so on. There was even a marsupial, an opossum. Two species of small horse have also been found in the Messel – all horses were small animals at this time. One of them was *Propaleotherium*, featured in the BBC's *Walking with Beasts*. The most famous fossil of all from the Messel shale is however *Darwinius*, known as Ida, a small mammal rather like a modern lemur.

****

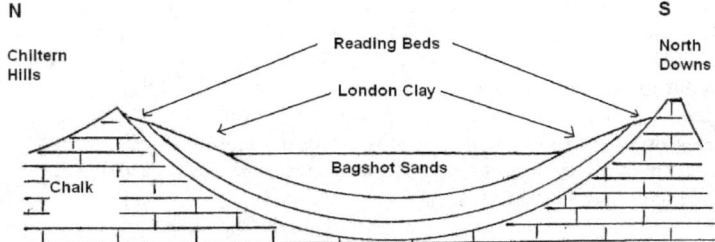

Cross-section of the geology west of London, between High Wycombe (north) and Aldershot (south). This is a perfect syncline with the newer rocks on top of the older.

Within England, the two great areas of Tertiary rocks are the London Basin and the Hampshire Basin. The rocks on the western side of this area tend to be sandy or gravelly (the Reading Beds), and formed in non-marine or coastal waters. On the eastern side, they are gluey clays (London Clay). The sandy facies tend to offer poor soil, either because they themselves lack fertility, or because they are overlain by flinty glacial gravel. The New Forest owes its continued existence to the fact that it grows on infertile land, as does Bagshot Heath (the Bagshot Sands). The poor quality of some of the river terrace gravels is due to the fact that they were what was left behind when the top of the Weald anticline eroded away. The Chalk dissolved, and all that remains of it is flinty gravel. There are other rocks, laid down in a shifting pattern as the sea transgressed and receded – the Barton Formation, the Bembridge Marls. All of these from the Reading Beds to the Bembridge Marls are Paleocene or Eocene. After the Eocene, England became an area of uplift and erosion, so that there are few rocks present from the next two epochs, the Oligocene and the Miocene. However in the south of Hampshire and the northern part of the Isle of Wight there is the Solent formation: silt, sand and shelly clay from the Oligocene. At Alum Bay on the Isle of Wight there is a beautiful and highly colourful exposure of rocks from the Chalk to the Bembridge Marls of the Oligocene. After the deposition of these rocks there is a gap of 25 million years before the next deposits, the Quaternary formations of East Anglia.

In northern England and Wales there are no Tertiary or Quaternary rocks (apart from boulder clay, reworked earlier material) and indeed it is estimated that a cover of 2 km (1.25 miles) of rock was stripped away after the end of the Cretaceous, about a fifth of it the Chalk itself.

There is a very important Tertiary formation in Britain which could scarcely be more different from the sedimentary rocks of the south-east. This is a volcanic province in the north-west of Scotland, centred on the Inner Hebrides, and particularly well represented on the islands of Skye, Mull and Staffa (the home of Fingal's Cave, organ pipes of basaltic hexagons which inspired Mendelssohn's Hebrides Overture). It appears again in Northern Ireland, where the basalt columns of the Giant's Causeway of Antrim are world-famous. These rocks are the legacy of the formation of the Atlantic, when Greenland drifted away from Scotland. What is now happening at the Mid-Atlantic ridge once happened in Scotland, but in fact the igneous geology is not just outpourings of basalt. There is a very rich texture of formations, including peridotites, ultrabasic rocks dredged from the depths. These represent the different phases of the period of vulcanism as the contents of the magma chamber changed. There is similar variety to be found in the modern igneous rocks of the Hawaiian Islands. The whole of Mull is really the remnant of an extinct volcano, and is the most complex igneous area in Britain. Much of Skye also represents exhumed volcanoes, since uplifted by about 2 km/6500 feet. One layer of ash from volcanic explosions in this area, dated to 52 million years ago, is found far away in the London Clay. This must represent an explosion of the same order as Santorini.

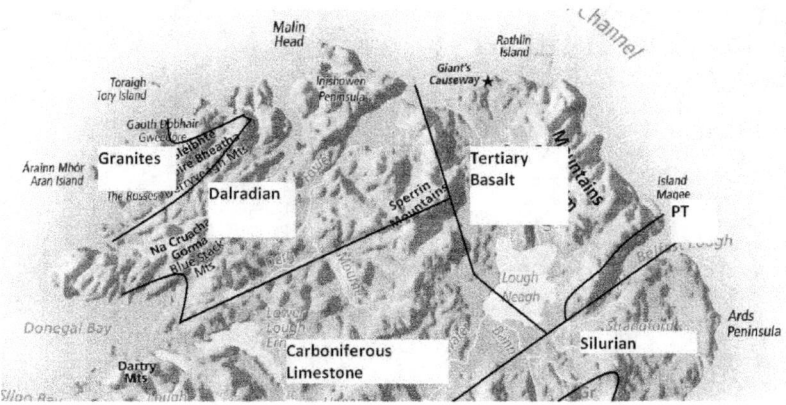

Geological provinces of northern Ireland. The Dalradian is composed of late Precambrian metamorphic rocks, a continuation of the same series in NE Scotland, including granites. The south-east has a province of Caledonian folding of Silurian and Ordovician rocks, again a continuation of the Southern Uplands of Scotland, with the same SW-NE trend. The centre of Ireland is composed of Upper Paleozoic rocks, mainly Carboniferous Limestone, seen at the base of this map. The north-east corner is the Tertiary volcanic province, once more a continuation of the formations on Skye and Mull. Finally there are small areas of Permo-Triassic sandstones as around the northern side of Belfast Lough.

Organ-pipe hexagonal basalt at the **Giant's Causeway**, Northern Ireland

On Skye are found two noted areas of intrusive rock, the Black and Red Cuillins. The Black ones are formed from gabbro, not a common rock elsewhere in Britain, but widespread here. It is a basic (mafic) formation, rich in dark pyroxene and green olivine. It cooled slowly so the crystals are visible to the naked eye. The Red Cuillins are the acidic (felsic) igneous equivalent, granites dominated by pink feldspars. In this area we also see the remains of calderas, volcanoes with their tops blown off, which indicated that the Tertiary eruptions were at times anything but peaceful. The whole country is also infused with igneous dykes, intruded into existing formations outwards from the volcanic centres. They swarm out from the Hebrides, running straight for miles, slicing through the country rock. The trend is NW-SE, at right-angles to the NE-SW Caledonian trend of the rest of northern Scotland. The rock in the dykes is dark, and without visible crystals – a form of dolerite, called quartz-dolerite. One dyke, the Cleveland dyke, extends 400 km/250 miles from the Hebrides, and can be traced across Cumberland and Durham. It is thought to have been formed in a period of days.

Tertiary igneous rocks also outcrop on the small island of Ailsa Craig in the Firth of Forth. This is a microgranite, famously used to make curling stones. (Microgranite is a felsic rock, intermediate in crystalline size between granite and rhyolite.)

A **red granite** of the type favoured on the high street, this one of Swedish rather than Scottish origin

****

Glaciation in the Antarctic began at the end of the Eocene as the continent shifted its position southwards over the pole. The Eocene ended with a rapid cooling of the earth from a mean temperature of 21 degrees Celsius to 16 degrees C. This was caused by feedback effects from the Antarctic, now sending waves of cold water and icebergs northwards into previously balmy seas, creating new patterns of circulation in the ocean. After 30 million years of boom, the mammals suffered their first mass extinction, but the main sufferers were in the marine ecosystems. It was slow process which nevertheless reduced the number of species by twenty percent – welcome to the Oligocene!

This epoch extends from about 35 million to 23 million years before the present. The name Oligocene comes from the Greek words meaning "few" and "new", simply meaning that few new mammals appeared in the Oligocene! It was a time of transition between the bygone world of the warm Eocene and the more modern-looking ecosystems of the next epoch, the Miocene. Major changes during the Oligocene included a global expansion of grasslands, and a shrinkage of tropical broad leaf forests to the equatorial belt.

Grass became a major plant in the scheme of life in the Oligocene and the next period, the Miocene. Thought to be a Tertiary evolution, grass pollen seeds have now been identified which push its origin back into the Mesozoic. Chewed grass has even turned up in the favorite rock of the professional geologist, dinosaur coprolite (fossilised dung). However only now did grass assume widespread importance. It has a remarkable property, utilized by you whenever you mow your lawn – its leaves grow from its concealed roots, not from the tips of its shoots, as happens with most plants. So it can be cropped without sustaining damage. It can make meadows, indeed whole savannahs, prairies and pampas, as no other plant can, and these grasslands soon came to sustain the major grazing animals as they continue to do today. During the Miocene grasslands were to take over whole regions of the continents, thriving in the cooler, drier conditions to become one of the most successful plant ecosystems ever to evolve. Grass is rather a tough plant, including silica in its structure, but animals soon found ways of taking advantage of it.

The newly open, grassy landscapes allowed animals to grow to larger sizes than they had earlier in the Tertiary. Many groups, such as horses, rhinoceroses and camels, evolved better runners during this time, adapting to the plains which were spreading as the Eocene rainforests receded. They were leaner, more muscular animals with longer legs and specialized feet with fewer toes. They developed specialized teeth for chewing the grass, and even found new ways of digesting grass, such as the multiple stomachs of ruminants.

South America was isolated from the other continents and evolved a quite distinct fauna during the Oligocene and also the Miocene, becoming a home to strange plants and animals. Elsewhere, brontotheres died out in the early Oligocene. In Asia, a group of running rhinos gave rise to the indricotheres, like *Indricotherium*, which were the largest land mammals ever to walk the Earth. They reached up to 8 metres (26 feet) tall and, as members of the rhinoceros family, were massively built pachyderms, weighing up to fifteen tonnes. This however does not approach the weight of the biggest dinosaurs.

"Dricky" the indricothere

During the Oligocene and the next epoch, the Miocene, Australia drifted through a much wetter climatic zone than its current position on the globe can offer, and the result was a rich, Amazonian-style flora and fauna. Some of this has been preserved in marvellous detail at Riversleigh in northern Queensland, one of those few sites in the world where every detail of the animals has been preserved. This revealed many types of previously undreamt-of marsupials, including one so odd it was christened the "Thingodont". There was even a meat-eating kangaroo.

Chapter 16 – Neogene

The Miocene epoch extends from 23 million years ago to 5.3 Million years ago. It was named by Sir Charles Lyell, author of *Principles of Geology*, after the Greek words for "less recent" because it has 18% fewer sea invertebrates than the Pliocene which follows it.

In terms of paleogeography, dramatic things were happening as the continents began to approach their present positions. The land bridge between South America and North America had not yet formed, but as South America approached the eastern subduction zone in the Pacific Ocean, the Andes began to rise. India by this time had run into the underbelly of Asia, creating enormous new mountain ranges, amongst them the Himalayas. At the end of the Oligocene, the African/Arabian plate ran into what is now Iran, cutting the Tethys Ocean in two. The eastern part of this then disappeared as Africa/Arabia collided with Turkey between 19 and 12 million years ago.

The subsequent uplift of mountains in the Mediterranean region and a global fall in sea levels combined to (temporarily) close the Straits of Gibraltar. This led to a drying up of the Mediterranean Sea (known as the Messinian salinity crisis) near the end of the Miocene. A thick layer of evaporites, six million years old, remains as evidence of this event, when the whole area must have been a desert. Remnants of Tethys remain today, to the east of the Mediterranean – the Black, Caspian and Aral Seas were all once part of it.

The climatic trend was towards increasing aridity caused primarily by global cooling, reducing the ability of the atmosphere to absorb moisture. Australia became drier as it entered a zone of low rainfall in the late Miocene. Further marked decreases in temperature during the middle Miocene at 15 million years ago reflect increased ice formation in Antarctica. It can be assumed that East Antarctica had a full ice cap by the mid-Miocene (15 million years ago). The Southern Ocean cooled partly due the formation of the Antarctic Circumpolar Current, as all the former parts of Gondwana apart from Antarctica itself were by now well clear of Antarctic waters. (The Greenland ice cap developed later, in the late Pliocene time, about 3 million years ago.)

Grasslands underwent a major expansion during the Miocene. The area of forest was reduced in response to a generally cooler and drier climate. Between 7 and 6 million years ago, there occurred a sudden expansion of grasses. Atmospheric carbon dioxide levels underwent a strange and unexplained drop to only about one-sixth of modern levels, but grasses adapted to evolve booster mechanisms to increase their intake. They also absorbed more silica, which they used to develop the cutting edges with which we are all familiar. These changes triggered a worldwide extinction of large herbivores, which had not evolved the special adaptations needed to crop and digest the new forms of grass, including tough lips!

The Miocene is sometimes known as the "Age of the Apes" due to the diversity of primates which evolved at this time, one of them the ancestor of *homo sapiens*. However there was also much diversity amongst elephant-type creatures, including the gomphotheres. Of these, *Gomphotherium* looks odd enough to those used only to elephants, a huge animal with four tusks instead of two. Another similar animal was *Deinotherium*, which had two downward-pointing tusks, designed for all the world as if to lever up buried treasure. In fact during the Cenozoic different groups of elephant-type animals – gomphotheres, mastodons, mammoths – have experimented with variations on the theme. Again, the cetaceans (whales, dolphins and porpoises) attained great diversity during the Miocene, with over 20 recognized genera in comparison to only six living genera today.

As Australia – or more correctly, the Australian plate, which includes New Guinea – moved into its modern position, a limited mixing of the Australian and Asian biota took place in the Malay archipelago, but the line between the two is still quite distinct in these island territories. In fact the demarcation zone is one of the most famous in biology as it was first remarked by Alfred Wallace, the co-founder of the theory of natural selection. He published a paper on the biogeography of the area in 1859, the year of the publication of Darwin's *Origins of the Species*. The line, ever since known as the Wallace Line, passes between the islands of Borneo and Sulawesi (Celebes), with Java, Sumatra, Malaysia and the Philippines on the north-western side of it, and New Guinea on the south-eastern side. The first contact between the plates is thought to date from 15 million years ago but transfers have been limited since.

****

It was during the Miocene that the main phase of the construction of the Alps took place. One famous place within them where the mechanics of mountain building were first worked out is the canton of Glarus in Switzerland. At the heart of the problem is a succession of rocks at a place called Lochstein. The bottom formation is called the Flysch, a dark grey slate. Unconformably above it lies a completely different rock, a conglomerate called the Verrucano. The pebbles within it have been elongated like sausages by metamorphic stretching. In between the two is a thin bed of limestone which has every appearance of having been squashed, in this deformed state known as mylonite. A horizontal crack runs straight through the middle of it. It was clear even in the nineteenth century that this sequence presented a puzzle, because the Flysch contains beautiful fossil fish from the Oligocene, drawn and described by none other than Louis Agassiz, who started out as a fish man. Some of them have been elongated to twice their normal length. This rock is now known to be 28 million years old. The Verrucano is know from outcrops beyond Glarus to be of Permian age, 250 million years old or more. So this is another version of the Moine Thrust problem.

Two things are obvious – firstly that the sequence is upside down, and secondly that this turnover happened after the Flysch was laid down. The Alpine geologists defined this as a nappe structure and in so doing, laid down the principles of mountain geology. The Verrucano has been thrust over the newer formation, its way lubricated by the limestone bed. This function is often performed by salt beds at other locations. The dispute was really about the number of nappes. At first it was thought that the structure was a double nappe, caused by pressure from both the north and the south throwing up two sets of folds. However the evidence on the ground eventually supported a single nappe pushed up from the south only – nearly all the structures in the rock face northwards. In any case the idea of two mighty pieces of crust moving together seemed unlikely. Even so it seemed incredible at first that such great masses of rocks could be thrust long distances at low angles.

The magnificent Alps: the Jungfrau, towering above Interlaken in Austria

****

In the "root zone" of the nappe, over the mountains in the Italian part of Switzerland, is a fault zone known as the Insubric Line. To the south lie the first rocks of the Adriatic and African plates which turned out to be doing the pushing – greenish schists, overlain in parts by Triassic limestones. There is no sign of intense folding. The jumbled mass begins north of the line. It is thought that there has been twenty kilometres (12 miles) of vertical movement along the line to bring the two sets of formations together at the surface today. The metamorphic rocks to the north have emerged from the hot depths of the earth. We know they have been there, because some of the contain eclogite, a garnet-bearing rock only formed very deep in the earth – "to exceed 70 miles", according to Arthur Holmes, an estimate not much changed today.

Modern methods can now date the mountain folding precisely. A dark, lustrous form of mica called biotite forms in metamorphic rocks in response to the temperature and pressure, and this contains a radioactive clock which was set at the time. The answer came as no surprise at 20 million years ago.

The effects of the Alpine orogeny were felt far beyond the Alps themselves – the Downs in England represent the outer flexing of the orogeny. The whole arch of the Carpathians in eastern Europe is a crumpling away from the plate crash. The Jura on the borders of France and Switzerland show a lower-grade, even type of folding, without the nappes of the Alps proper. As the African plate has advanced northwards, it has also pushed Spain into France, forcing up the Pyrenees and both crumpling and raising the level of Spain. Much the same thing happened on the eastern side of the Mediterranean in Anatolia. The Caucasus mountains of south-west Asia and the Atlas Mountains of North Africa have a similar origin.

So it was that the mechanics of mountain building were deciphered, though the driving force behind it remained obscure until the second half of the twentieth century. Eduard Suess, writing before the discovery of radioactivity, thought that the earth was cooling and so shrinking. This would necessarily lead to a contraction of the existing crust, which would cause mountains to be thrown up as the segments of it struggled to fit into a smaller space. For the time, this seemed a reasonable explanation, but it is now known that the earth is neither cooling nor contracting.

The Alps we see today are rather poor in granites, probably because these are new mountains, not yet worn down to their roots. They do include some granites, notably the Mont Blanc massif, but these are relics from Hercynian mountains caught up in the new earth movements.

****

The Miocene was a busy time for mountain building, because this is when the main phase of the construction of the Himalayas took place. In 2004 I decided I had better check them out myself. Christine and I went with our favorite tour group, Explore. Nothing can quite prepare you for the tumult and chaos of a place like Delhi, the lunatic driving, the dust, the colour, the beggars, the motorized rickshaws. "Chaos" is a word which could have been invented just to describe India. So, we joined the Explore group, then picked up a flight into the Himalayas. We were headed for Leh in Ladakh, beyond the first range of really high mountains, where the river Indus cuts its way through the mountains heading west, before turning southwards to its eventual outlet in the Indian Ocean on the coast of Pakistan. This far into the Himalayas, the river creates a linear oasis, because the mountains themselves are deserts – all the rain has been washed out at lower altitudes.

Leh itself stood at over 3,000 metres (10,000 feet), the mountains around it 7400 metres (24,000 feet) and more. It was rather like Colorado, but 10,000 feet higher with a lot less air to breathe! The views were stupendous, but it was hard enough to sleep in Leh itself. Many dogs could be seen sleeping on the pavements during the day, then they barked all night. I don't know why the locals put up with it. It's not as if it's their religion – they are Buddhists up there, and there are dozens of monuments and monasteries, some impressive, all idolatrous. Believe it or not, one of our tour party, a woman teacher in her thirties, English but working in Switzerland, was afraid of heights! She slowed things right down when we were out trekking.

The shortage of air was bad enough in Leh, but it got worse. We eventually left Ladakh to make our way back down to the plain of the Ganges, heading east then south. The highest pass, Tanglangla, was exactly 5410 metres (17,582 feet) up. I got out here and lit a small cigar, then immediately turned a horrible blue colour. The tour guide looked seriously alarmed and dug out an emergency oxygen kit, but I smoked the cigar anyway. The views further down were spectacular, but our problems were hardly over. Above a place called Manali, the road had collapsed in the monsoon rain (apparently this happens quite often), and no way round. We had to walk the last few miles down an extremely dangerous, muddy and slippery slope. Even wearing proper hiking boots, Christine fell over three times, and split her lip open. Still we made it after a couple of uncomfortable hours.

Geology in action – a landslide blocks the Manali Road

Even then it was a long, long ride down through Himachal Pradesh, but before we hit the plain, we started to go back up – we were going to McLeod Ganj, near Dharamshala, home of the exiled Tibetan Dalai Lama. Situated at 1845 metres (6000 feet), apparently almost permanently wet and foggy, this place is about as far from the popular conception of India as can be imagined. It was full of tacky tourist shops and rubbish on the streets. All I can say is that, as a colonial hill station, it must have been very agreeable after the heat of the plain, and there were plenty of signs of the old English settlement, including a church and a graveyard. But don't bother to go there now. From here we finally descended into the Punjab at Amritsar. The Golden Temple, holiest place of the Sikhs, is plagued by beggars, but we went in. We were a bit afraid of upsetting the sensibilities of the real worshippers as we got into the centre, but a local girl egged us on!

The main drainage of this central section of the high Himalayas is via three river systems, the Indus, the Brahmaputra and the Ganges. This last river takes the easy route, draining the southern slopes and picking up tributaries as it does so, more of them flowing in from the south when it reaches the Indo-Gangetic plain. Both the Indus and the Brahmaputra start behind the main range, and then have to cut a way through to the sea, in the manner of the Colorado River, maintaining their course as the land rose around them. These are examples of antecedent drainage where a river exists before the structure and manages to maintain its course by cutting through the new structure. The gorges of the Brahmaputra are apparently particularly spectacular. Behind the Himalayas the river is known by its Tibetan name, the Tsangpo. Turning southwards through the gorges, the great river flows through Assam, eventually to join the Ganges in one great delta.

The Himalayas are certainly the highest mountains in the world, but some have questioned why they are not even higher. Detailed ground geology has shown that it is not simply a question of the Indian plate ploughing beneath the Asian plate. At the top of the range, geologists have found a series of normal faults, which are causing the tops of the mountains to slip downwards (northwards). Underneath them at much lower levels are thrust faults which are moving the lower parts of the crust upwards (southwards). Both types of faults have been active at the same time. Thrust faults are a normal part of the mountain-building process, but the normal faults were not expected. It appears that the Himalayas are collapsing at the top.

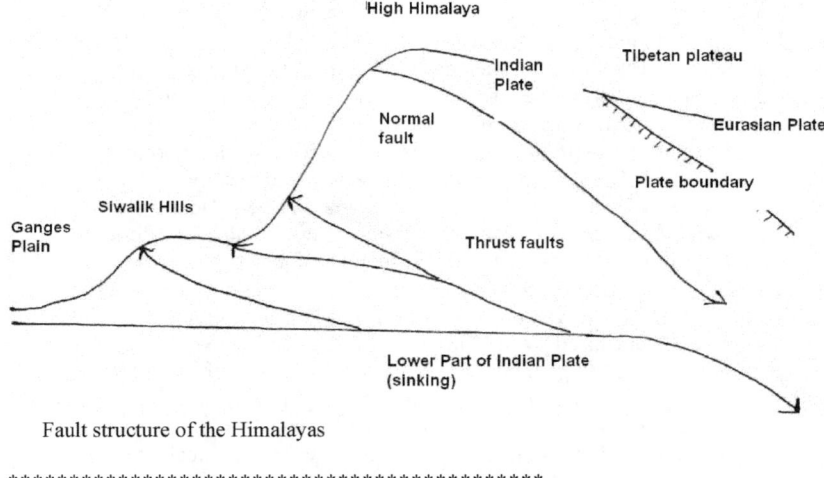

Fault structure of the Himalayas

****

The Pliocene epoch extends from 5.3 million years ago to 2.58 million years ago. It is the second and youngest epoch of the Neogene Period in the Cenozoic Era. Once again it was named by Sir Charles Lyell, from the Greek words meaning "continuation of the recent", referring to the essentially modern marine mollusc faunas.

During the Pliocene epoch the climate became cooler and drier. The global average temperature in the mid-Pliocene was 2-3°C higher than today, and global sea level was 25 m (80 feet) higher. But something momentous happened three million years ago – North and South America became joined together at the Isthmus of Panama. This changed the circulation of the oceans in a fundamental way in that the current flowing through the old gap now flowed northwards. Ocean currents are by far the most important redistributors of heat on the surface of the earth. They act like vast storage heaters which serve to dampen the excesses of winter. The general surface movement is to carry the equatorial and tropical waters to the poles, and this continued to happen in the North Atlantic, as the Caribbean waters became the Gulf Stream, heading north and east. The presence of this enormous flow of warm water shields western Europe from the worst of the cold, raising the temperature by a hefty 6 degrees Celsius over comparable latitudes which do not receive a benign ocean current – for example Labrador – but even this circulation was not sufficient to hold the ice sheets away for long. Just the same, it may not be a coincidence that the Ice Ages started so soon after North and South America became conjoined. The Atlantic was now cut off from supplies of warm equatorial waters from the Pacific at just the time when large influxes of cold water were flowing into it from the poles.

The main discussion of the causes of the Ice Ages is deferred until the next epoch, the Pleistocene, though there is evidence that glaciation had started in the less hospitable parts of the Northern Hemisphere in the onset of extensive glaciation over Greenland in the late Pliocene around 3 million years ago. The formation of an Arctic ice cap is signaled by an abrupt shift in oxygen isotope ratios and ice-rafted cobbles in the North Atlantic and North Pacific oceanic sediments. The fall in sea level caused by the formation of land ice in the polar regions exposed the land-bridge between Alaska and Asia. Note that sea ice has no effect on sea level as it displaces its own weight in water.

The change to a cooler, dry climate had a considerable impact on Pliocene vegetation, reducing the coverage of tropical species. Deciduous forests proliferated, coniferous forests and tundra covered much of the north, and grasslands spread on all continents (except Antarctica, of course!) Tropical forests were limited to a tight band around the equator, and in addition to dry savannahs, deserts extended in Asia and Africa.

By this stage, South America had developed a marvellous array of unusual creatures. Perhaps the strangest was the glyptodont, a tank-sized armadillo with a great, spiked wrecking ball at the end of its tail. Then there was the giant ground sloth, *Megatherium*, 6 metres (20 feet) long from head to tail, and rodents the size of bears called (in translation) "terrible mice". When skeletons of the great-boned *Megatherium* first appeared in the museums of Europe towards the end of the eighteenth century, before dinosaurs were known, they caused a sensation.

The linkage of North and South America made possible a great interchange of flora and fauna, and brought a partial end to South America's distinctive native fauna, isolated from North America ever since the days of Pangaea millions of years previously. It is notable that this could not have happened to the same extent if the flora had been the hot, sticky and almost impenetrable jungles of Darien that were encountered by the Spanish Conquistadores. The migrations must have taken place in a drier phase. Ground sloths including *Megatherium*, glyptodonts, armadillos and porcupines headed north across the isthmus, as horses, dogs, deer, pumas, bears and elephants headed the other way. Also moving south was the North American camel, ancestor of the llama. The top predators of the rapidly expanding pampas, the terror birds, were suddenly replaced by an even more fearsome animal from the north – *Smilodon*, the sabre-toothed cat. This animal was also widespread in Eurasia. Alligators and crocodiles died out in Europe as the climate cooled down.

In Australia, the marsupials remained the dominant mammals, including the huge *Diprotodon* or giant wombat, thought to weigh over two tonnes. This animal was preyed upon by *Thylacoleo*, a marsupial lion!

****

The only place in Britain with any appreciable Pliocene deposits is East Anglia, where they are found in a broad band of country extending from south of Ipswich to the north coast of Norfolk. There is one small patch of Coralline Crag, which is neither a of coralline origin nor a crag, but a shelly gravel. The main formation is a poorly consolidated shelly sand known as the Red Crags, easily eroded along the coast. These include a wide range of fossils, both marine and terrestrial. Amongst these are elephant's teeth! Larger than a loaf, heavily ridged on top and weighty, these are not uncommon. To the west of the Red Crags lies the Chalk, in places completely covered by Pliocene or Pleistocene sands and gravels. One such area is the extensive Breckland around Thetford, which has proved useless for agriculture and is now the site of a coniferous forest.

There is widespread evidence from abandoned strandlines and hillside bevelling that Britain was inundated to a depth of 180 (585 feet) at the end of the Pliocene and the start of the Quaternary. This is known as the Calabrian transgression.

Another unstable coastline – limestone cliffs on the Greek island of **Xante**. Greece is the earthquake capital of the world and these cliffs look as if they have been uplifted from the sea in the not-too-distant past.

Chapter 17 – Quaternary

The Pleistocene is the epoch from 2.58 million years ago to 11,500 years ago, recently pushed back in time so as to include all the repeated glaciations of the northern hemisphere. Its name is derived from the Greek for "most new". The end of the Pleistocene corresponds with the end of the last glacial period. It also corresponds with the end of the Paleolithic age used in archaeology.

As far as the onset of the Ice Age is concerned, the distribution of the continents at this time is critical, but there is an important difference between the northern and southern hemispheres. By seven million years ago, Antarctica had moved plumb over the south pole, and glaciation had begun in East Antarctica well before that – as early as 35 million years ago on some estimates. No ocean current could penetrate the area to warm it up. However the north pole is almost the negative image of the south. Instead of a continent surrounded by oceans, there is an ocean surrounded by continents. There is only one real entrance to this ocean, through the gap between Greenland and Norway. The back door – the Bering Strait between Siberia and Alaska, only 85 kilometres (58 miles) wide now – was sometimes completely closed when sea levels fell as a result of the formation of land ice. This would have prevented ANY cold Arctic water escaping into the Pacific, so that in turn warm Atlantic water could not flow through the Arctic. If you leave the front and back door of your house open, a strong breeze will blow straight through. If you close the back door the breeze will drop to almost nothing. Warm water was blocked from the Arctic by cold water unable to move out of the back door. This effect still operates today. The Gulf Stream flows northwards to Norway, evaporating and becoming denser all the way until it sinks as it is heavy with salt – but if it could push the cold Arctic water out through the back door, it would carry on northwards, and there would be no Arctic ice sheet, though there would still be one on Greenland. Only a small fraction of the massive amount of heat transferred by the Gulf Stream would be sufficient to melt the Arctic ice cap.

In fact the complete closure of the Bering Strait is known to have taken place as several points in the ice ages, forming a land area known as Beringia. The current seas are so shallow that it is thought that this covered a very extensive area, 1000 kilometres (600 miles) from north to south. The emergence of such a complete barrier to oceanic currents would have had a feedback effect, reinforcing the ice age once it emerged from the seas. However this land bridge does not have to be 1000 kilometres (600 miles) wide to stop the currents – it need only be a yard wide! It is flooded now, and we are in an interglacial. It is known to have dried out in glacial periods. As it is in fact a shallow area of continental shelf, it could be flooded quite quickly, within a decade.

So it is likely that the increasing stranglehold on the Arctic by converging land masses created the ice ages in the northern hemisphere. However, biogeographical evidence demonstrates previous connections between North America and Asia. Similar dinosaur fossils occur both in Asia and in North America. For example, relatives of *Triceratops* and even *Tyrannosaurus rex* came to North America from Asia. There is also evidence of an interchange of mammal species around 20 million years ago. Some, like the sabre-toothed cats, have a wide geographical range in Europe, Africa, Asia, and North America. There is also a marked similarity between the vegetation of North America and that of Eurasia. The farther north one goes, the greater the similarity. The difference in the Pleistocene Arctic may be that Greenland had moved into a position to narrow the ring of land around the Arctic.

Fluctuations in the amount of heat received from the sun known as Milankovitch cycles are nowadays thought to have caused the massive swings in temperature which occurred during the Ice Age itself. Milankovitch was a Serbian civil engineer and mathematician who worked out his theory whilst in internment during the First World War. There are three aspects to it: the precession of the equinoxes, the variable eccentricity of the orbit of the earth around the sun, and the axial tilt of the earth.

The precession of the equinoxes refers to the earth's slow movement backwards along its own orbit, so that for example on 21 March this year we would be looking at the sun from the very opposite end of the orbit to that on 21 March 13,000 years previously. This phenomenon also causes the stars we see at night slowly to rotate, as in the phrase "the dawning of the Age of Aquarius" – following the previous astronomical age, Pisces (the sequence runs backwards). The axis of the earth completes one full cycle of precession approximately every 26,000 years. At the same time the elliptical orbit rotates more slowly, on a cycle of 100,000 years. The orbit is close to a full circle, being only slightly elliptical, but the point of maximum distance from the sun in say the northern hemisphere summer does vary over time, due to the influence of the gravity of Jupiter and Saturn. The combined effect of the precession of the equinoxes and the variable eccentricity of the orbit is a 22,000-year cycle between the maximum and minimum insolation of either hemisphere. In addition, the angle of the earth's rotational axis oscillates between 22.1 and 24.5 degrees, on a 41,000-year cycle. It is currently 23.44 degrees and decreasing.

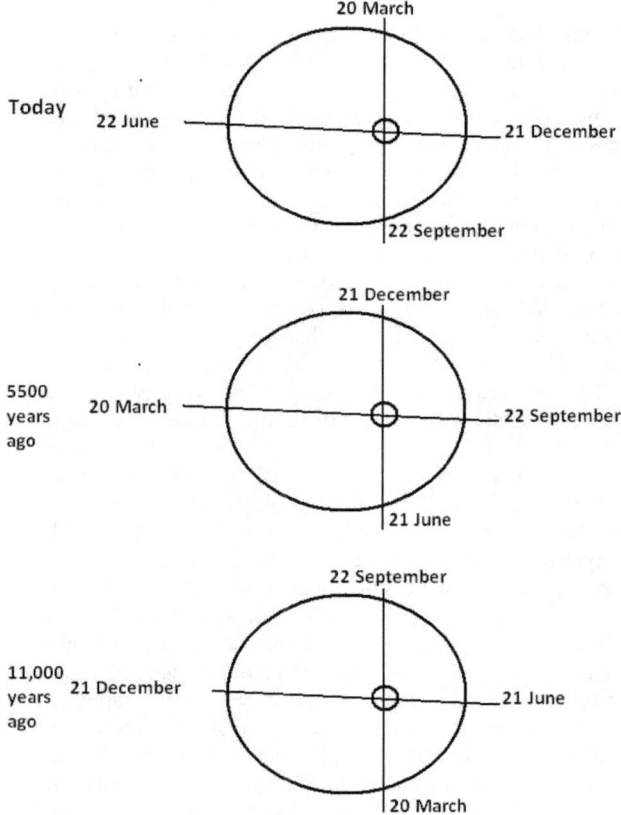

The combined effects of precession and the rotation of the Earth's orbit. Today winter in the Northern Hemisphere occurs when the earth is at its closest to the sun. 11,000 years ago it occurred when the Earth was at the far end of its orbit. The visible effect is that the first set of stars we see each night gradually changes.

****

The cycles have little or no effect on the total annual budget of heat from the sun (Tim Flannery quotes a figure of less than 0.1 percent, page 42). However they affect the distribution of heat over the planet and so the climate of the earth. For example in years when the three factors combine to give maximum insolation to the southern hemisphere, a proportion of that insolation will be reflected straight back out into space by the Antarctic ice cap, and so be lost. At the same time, the northern hemisphere would have a long run of the type of very cold summers needed to form an Arctic ice sheet.

There has been an increasingly good match between ice age temperatures dated from thorium isotopes taken from oceanic cores and the Milankovitch cycles, which had been abandoned as of little interest by the 1970s. There are good matches to the cycle in the peak glaciations at 100,000, 42,000 and 21,000 years ago. However, it is as well to remember that the cycles have no claim to cause the ice ages, but only to affect the cycles within the ice age. The ice age was brought about by a steady decline in world temperatures over 50 million years from the Paleocene, and that fall can probably be put down to the movements of the continents in the polar areas, as stated above. The Milankovitch cycles were there before the ice age ever began.

There is an important feedback effect which has influenced the ebb and flow of the ice sheets. Ice has a very high albedo, or ability to reflect sunlight back into space. The more the ice spreads, the more light it reflects away, the colder it becomes, the more the ice spreads. If the sun's rays hit dark oceanic water, the heat they carry is retained, so the same process operates in reverse – once the ice starts to recede, the more heat is retained, the faster it recedes. Studies show that when the ice sheets have started to melt, they have shrunk very rapidly, but freezing can be equally rapid. Studies of ice cores from Greenland show that an extreme oscillation from balmy to freezing conditions in England could take place in only ten years.

The enormous amount of water locked up in the ice meant that the world was a much drier place. The fall in the content of water vapour in the atmosphere was probably a contributory factor in maintaining glaciated conditions, as water vapour is an even more important greenhouse gas than carbon dioxide.

In recent years, another form of feedback has emerged from cores extracted from the Antarctic ice. These and other sources have shown that the carbon dioxide content of the atmosphere was very much lower at the peak of the last glaciation, 18,000 years ago – only 200 parts per million, as opposed to 280 ppm in the pre-industrial age, and 380 ppm today. Of course this may well have reduced the warming, greenhouse effect of carbon dioxide, but was it a cause of the ice age, or an effect? There is strong evidence that is was an effect. The ice ages were very dry, and dusty, iron-rich winds blew off Patagonia across the Southern Ocean, greatly enhancing the supply of iron which phytoplankton need to grow. The result was massive blooms, reflected in the fossil record, which would have drawn down an estimated two billion tonnes of carbon dioxide from ocean, which would have then replenished itself with dissolved carbon dioxide from the atmosphere. So once an ice age got going, it would be greatly intensified by this effect, even if it is doubtful that it alone could have accounted for all the drop in carbon dioxide levels. Nevertheless the levels of carbon dioxide in the atmosphere did rise and fall in lockstep with the global temperature.

The same kind of result has come from measuring the level of dimethyl sulphide in the ice cores. This chemical is produced by marine algae and is thought to play a major role in the seeding of clouds. The measurements show that the output of DMS was nearly five time greater in an ice age than otherwise. The algae must have been doing well, and moreover, there must have been a lot of cloud! – reflecting away the sun.

****

Pleistocene climate was marked by repeated glacial cycles where continental glaciers pushed to the 40th parallel in some places. It is estimated that, at maximum glacial extent, 30% of the Earth's surface was covered by ice. In addition, a zone of permafrost stretched southward from the edge of the glacial sheets. It has long been recognised that there were separate surges of glaciation within the ice Age. Forty years ago, it was thought that there had been four main glaciations, called Gunz, Mindel, Riss and Wurm (known to my old geography teacher as the Great Midland Railway of Wales). Paleoclimatic studies based on ice and ocean cores now show that the ice has repeatedly come and gone in cycles of a hundred thousand years or so, perhaps fifty times, and that even in the depths of glaciation, there have been sudden spikes of warmth. The longer interglacials featured times when the climate was a good deal more equable than it is now.

Each glacial advance tied up huge volumes of water in continental ice sheets 1,500 to 3,000 metres (4,900–9,800 feet) thick, resulting in temporary sea level drops of 100 metres (325 ft) or more over the entire surface of the Earth. Some put this figure at 120 metres (390 feet), uncovering an area of land about the size of Africa, much of it in warm places in south-east Asia. Note however that to reach such extreme falls in sea level we would need to go back to a starting figure from millions of years ago, before the Antarctic ice sheet formed. Conversely during interglacial times, such as at present, drowned coastlines resulted from rising sea levels.

Columnar basalts from the **Upper Loire Valley, France, near Le Puy-en-Velay.** High areas even in southern France would have carried ice caps, like the ice caps of Iceland today.

Ice cores from Greenland which can go back 400,000 years have shown temperature changes of an amazing six degrees Celsius within a decade! – enough to start or end an ice age. There has been much speculation about what the cause of such rapid changes in temperature could be. The prime suspect is the closure or reopening of the Gulf Stream, the so-called "Atlantic Conveyor", possibly because of the release of melted ice-water into it, reducing its salinity, and so stopping it from sinking. The sudden breaching or closure of the Bering Strait could also have changed the climate quickly. Another possibility is that the climate of the earth is simply much more volatile during an ice age than in an interglacial – indeed, that appears to be the case – as if there are two stable states, frozen and interglacial, and the climate hops between one and the other with not much in between. (Even Hollywood has shown an interest in this, in the 2004 film *The Day After Tomorrow*, when the Atlantic Conveyor shuts down.)

The effects of glaciation were global. Antarctica and Greenland have been ice-bound throughout the Pleistocene. The Andes contained a series of ice caps including a large one in the south, the Patagonian. There were glaciers in New Zealand and Tasmania. Mountainous areas of Africa including East Africa, Ethiopia and the Atlas Mountains all had ice caps. Continental-sized ice sheets covered the American north-west and north-east. In Europe, the Scandinavian ice sheet extended right across the North Sea to cover most of Britain. The Alpine ice sheet extended far from the mountains. Scattered domes stretched across Siberia and the Arctic shelf. However some northern areas of Russia, and also "Beringia", escaped glaciation simply because there was so little precipitation.

It takes a lot of life to reduce carbon dioxide to below 200 parts per million, so the ice ages were by no means bad news for life on the planet as a whole. Aside from algal blooms, they produced a surprisingly successful sub-tundra fauna, given that nothing much apart from various types of reindeer and elk seem to prosper in similar regions today. In fact a large body size is efficient at conserving heat. Fur, which must have been a nuisance during the scorching Eocene, suddenly came in very handy for the mammals. The most notable animals developed adaptations to the cold, and thrived in the conditions – woolly mammoths, rhinoceros and bison. Animals which were larger than their founding stock include the cave bears, the European lion, the polar bear and the Irish elk (not surprisingly known as *Megaloceros giganteus* – gigantic big horn). There must have been something about the conditions which favoured such massive headgear, because the mammoth had the largest known elephantine tusks, arranged in spiral form and measuring up to 5.5 metres (18 feet). Predators too developed adaptations, as in the white fur of the Arctic fox and polar bear.

****

Not least amongst the achievements of Pliocene fauna was the evolution of early hominids, though most of the evidence then comes from the Pleistocene. Footprints of two hominids were found in 1976 at Laetoli in Tanzania in volcanic sediments which can be precisely dated to 3.6 million years ago. These indicate an upright stance, confirming the long-held view that bipedalism was the earliest of the adaptations that turned an ape into a person, preceding, for example, large brain size. In 1974, along came Lucy. A party led by Don Johanson found forty percent of a full skeleton, an unprecedented amount for such old bones. She is now classed as *Australopithecus afarensis* and dated to 3.1 million years ago – younger than the footprints. She is quite a small figure, and not especially robust either. If Neanderthal man came and sat next to you on a bus, you would get off the bus as fast as possible. If it was Lucy, you might rummage in your bag for a banana to give her.

The first skeletons to be classed as people – *Homo habilis*, handy man, maker of stone tools – date from about two million years ago. Stone tool assemblages date back to 2.5 million years ago, the design of the tools then not changing for another million years. The next important hominid is *Homo erectus*, whose brain capacity lies half-way between that of Lucy and modern man. He is known from fossils found in Ethiopia dating to 1.7 million years ago, but he spread round the world. Elsewhere he is known as Java Man and Peking Man. He seems to have disappeared about 1.3 million years ago, possibly the victim of a particularly vicious glaciation.

Searching for the bones of ancient hominids is a laborious process but by now a number of other types have been identified, with bewildering and ever-changing names. The fact remains, however, that all the early hominid bones ever found could probably fit into the back of an estate car. It is apparent, however, that one feature evolved above all others in the period between 2.5 and 1.5 million years ago – the brain – which increased fourfold in size in this period.

The first modern man, *Homo sapiens* himself, appeared around 150,000 years ago. He was distinguished from all his predecessors by his marvellous dexterity – who ever heard of a chimpanzee playing a violin? – and also of course by his ability to talk. Most of the expansion in the size of his brain was connected with these two attributes – for example, the ability to speak requires very fine musculature around the mouth, which is controlled from the brain. Neanderthal man probably split off from the main human branch about 600,000 years ago. He is especially associated with the Ice Ages and had several cold-climate adaptations including a squat, sturdy build (like an Eskimo) and a large nose to warm the air.

One interesting but for some reason neglected theory about *Homo sapiens* is that some time very early in his evolution, going back perhaps five million years, he underwent an aquatic phase – the "Aquatic Ape". This idea, promulgated enthusiastically by Elaine Morgan, would explain some otherwise puzzling features – hairlessness and the layer of subcutaneous fat, both like a seal; the direction of what body hair remains, aligned to the flow of water; the downward-pointing nose; bipedalism itself, suitable for swimming as well as walking and running; the ability to hold the breath, which is rare amongst terrestrial mammals, but obviously needed for time spent under water; and the "drowning reflex" where every body function but the operation of the brain is shut down in sequence during the process of drowning. It's an excellent case.

Yet another type of human is also now know, from a single find of fingers and teeth in a cave in Russia. The is known as Denisovan man. Although the remains are 40,000 years old, DNA analysis has been successfully carried out, indicating that this was NOT *homo sapiens*. However some Denisovan genes are thought to have survived in modern people by earlier interbreeding, particularly one which enables the modern Sherpas to thrive at high altitudes.

****

At some stage in the Ice Age, the English Channel was formed, probably the most important geological event in world history. Until a maximum of 400,000 years ago, England was connected to France by a ridge of chalk between Kent and Artois. Even as sea levels rose and fell with the coming and going of glaciations and interglacials, this ridge held, and early man could walk into England if he so chose. The drainage from the English plain was north of the ridge, south of "Doggerland", now an area of shallows in the middle of the North Sea, to join and extended Rhine in a great estuary or delta. In ice age conditions the whole of the southern North Sea would have been a mass of semi-frozen fens and islets, rather in the manner of the mouth of the

Lena in Siberia today. Some take the view that this was still the position 200,000 years ago! However, at some point a great periglacial lake – formed by the drainage of these same rivers, the proto-Thames, Rhine and also the Maas – built up to such a height that the Kent-Artois ridge was breached. Undersea scans reveal a valley at the bottom of the Channel, about 10 kilometres (7 miles) wide and 50 metres (160 feet) deep, with almost vertical sides. It is believed that this was cut when the ridge was breached. From that point onwards, the drainage of the Rhine, Thames and also the Somme was down the Channel, not out into the North Sea, and that pattern would re-emerge if sea levels were to fall again. At times of high glaciation, this was still only a river. Only when the ice melted during interglacials – for the last time, no more than 8,000 years ago – did it form a sea lane. Its existence was eventually to lead to Britain developing the finest navy in the world, and the country was able to establish overseas colonies in North America, India, Australia and elsewhere. This in turn is now leading to the establishment of English as the world's *lingua franca*. In 1940, Winston Churchill flew to France to try to rally the French leaders in the face of the blitzkrieg, which was about to knock them out of the Second World War. One of them, General Weygand, remarked ruefully that "England has rather a good anti-tank ditch"! Geology has formed England, and with a little help from the locals, kept it independent.

South of the ice sheets large lakes accumulated because outlets were blocked and the cooler air slowed evaporation. When the Laurentide ice sheet retreated, north central North America was totally covered by Lake Agassiz. Deserts on the other hand were drier and more extensive. Increases in aeolian deposits – wind-born dust, forming a new land cover or loess – are recorded from glacial periods in China and Europe.

The severe climatic changes during the ice age had major impacts on the fauna and flora. With each advance of the ice, large areas of the continents became totally depopulated, and plants and animals retreating southward in front of the advancing ice sheets faced tremendous stress. The earlier ice ages lasted for about 40,000 years, but the later ones for as long as 100,000 years, putting great pressure on animal populations. From about 300,000 years ago, formerly common animals started to become extinct. The 100,000-year cycle of the variation in the ellipse of the orbit of the earth maps well onto this later phase of the ice ages, whereas before 700,000 years ago, the beat seems to have been led by the 41,000-year cycle of the tilt of the axis of the earth.

In recent years, the local names for glaciations and interglacials have been replaced in some texts by an international system based on the ratio of the isotope of oxygen-18 found in marine deposits of foraminifera shells. This ratio increases during glaciations as the lighter oxygen-16 is evaporated and ends up on the land as ice. The new divisions are known as Oxygen Isotope Stages (OIS) or Marine Isotope Stages (MIS). The most recent history of these is:

OIS1 - to 12,000 years ago – interglacial
OIS2 - 12,000 to 30,000 years ago – full-scale glaciation (Devensian)
OIS3 - 30,000 to 60,000 years ago – interstadial, in and out of deep freeze
OIS4 - 60,000-75,000 years ago – full glaciation
OIS5 - 75,000 to 130,000 years ago – long interglacial (Ipswichian)
OIS6 - 130,000 to 195,000 years ago – full glaciation (Wolstonian)
OIS7 - 195,000 to 255,000 years ago – interglacial
OIS8 - 255,000 to 290,000 years ago – full glaciation (Wolstonian)

In some classifications the entire period between 130,000 years ago and 12,000 years ago is classified as the Devensian ice age, but the evidence is that it was far from being a continuous glaciation – in fact an ice sheet is thought to have been present on Britain for only a tenth of that time. The term Wolstonian is used for the full period between 290,000 and 130,000 years ago. Well-known earlier stages are the Hoxnian interglacial (424,000-374,00 years ago, OIS 11) and the Anglian glaciation (478,000-424,000 years ago, OIS 12).

In Britain, maximum glaciation (down to the Thames Valley) was achieved in OIS6. The last glaciation, OIS2, did not extend quite so far southwards. It is thought that Britain did not suffer a full glaciation until OIS8, starting a comparatively recent 290,000 years ago. It is thought that the Scottish Highlands were probably glaciated well before that, but there is no actual evidence on the ground. This is of course a problem with the ice ages – a new ice age destroys much of the evidence from the previous one, and estimates for the number of ice advances vary widely. It is certainly rather alarming that, far from receding, the ice ages are, in their overall pattern, intensifying.

As the climate fluctuated wildly between ice sheets and tundra on the one hand, and subtropical interglacials on the other, anything can turn up in the fossil record. The Ipswichian Interglacial from 130,000 years ago is well recorded with finds of tropical animals – hyenas, hippos and elephants – turning up along the Thames Valley. One casualty of this interglacial may have been *Smilodon*, the sabre-toothed cat, which disappears from the fossil record about 100,000 years ago. The Ipswichian was so warm that forests would have invaded the open grasslands which were its hunting grounds. By contrast, at the ice maximum, England was connected to the continent, the Thames was a tributary of the Rhine and the North Sea was a mass of fens and islets. Scotland was connected to Ulster, and the Isle of Man and Anglesey were not islands.

Hominids established themselves in Britain during some interglacials, when they at times shared the land with such exotic creatures as hippos, lions and hyenas. The earliest evidence, stone tool assemblages, comes from Pakefield, near Lowestoft in Suffolk, and is thought to date from 700,000 years ago. The first actual hominid bone found so far comes from Boxgrove Man, who lived in Sussex half a million years ago. Only a single tibia has been found, thought to be a bone from an early hominid, *Homo heidelbergensis*. At Swanscombe in Kent, most of a skull was found, dating from the Hoxnian interglacial and dated to 400,000 years ago (probably *Homo erectus* or *heidelbergensis* as this is much too early for *Homo sapiens*). After that, until 60,000 years ago, the evidence for the presence of hominids is sparse, and lost sight of entirely for long periods, despite the known mild interglacial conditions which occurred within the ice ages.

Neanderthal man, first identified in Germany, also lived in Britain. His bones have been identified in a cave at Pontnewydd in North Wales, dated to 240,000 years ago. Neanderthal man was short, stocky and robust, with bandy legs and a figure in general adapted to ice-age conditions. There is no evidence of any emergence from Africa – he is thought to have evolved on the fringes of the northern ice sheets from resident hominids. After 200,000 years ago, until 60,000 years ago, there is little if any evidence of hominids in Britain. This period includes the entire last great interglacial, the Ipswichian, when conditions were perfect for such a hominid – but for some reason, he appears not to have reached here. In fact Neanderthal bones are rare in Britain, but their distinctive tool assemblages are quite widespread across the south of England.

Homo sapiens, our own species which had entirely replaced Neanderthal man by 30,000 years ago, was a much more gracile type who had evolved in the heat of the Africa. During the peak of the ice ages modern man retreated like a sensible fellow to the south of France, where as Cro-Magnon man he drew cave paintings 30,000 years ago. From this late ice age one famous skeleton is known, the Red Lady of

Paviland, found in a cave on the Gower Peninsula in South Wales and identified by the geologist William Buckland in 1823. Dated to 26,000 years ago, the skeleton (headless, and now thought to be a man) was stained with red ochre, and decorated with periwinkle shells and ivory ornaments.

****

About 15,000 years ago, the ice started to melt for the last time – for now – and within a few thousand years, most of it had disappeared from the northern hemisphere. Warming continued up to a maximum temperature at about 6,000 years ago, when elephant and giraffe grazed in the middle of the Sahara Desert. The post-glacial world is represented in geology as the Holocene epoch which began at the end of the Pleistocene 11,500 years ago and continues to the present.

Ice melt caused world sea levels to rise about 35 m (114 ft) in the early part of the Holocene. The maximum fall was 100 metres (325 feet) and even today there is a lot of land ice left in the Antarctic and Greenland – enough to raise sea levels by another 65 metres (211 feet). In addition, many areas above about 40 degrees north latitude had been depressed by the weight of the Pleistocene glaciers and rose as much as 180 m (600 ft) due to post-glacial rebound over the late Pleistocene and Holocene, and are still rising today. An extreme uplift of 270 metres (900 feet) has been measured in the northern part of the Gulf of Bothnia (Finland), which continues at the rate of about a centimetre a year.

A major extinction event of large mammals began late in the Pleistocene and continued into the Holocene. Included in it were mammoths, mastodons, South American ground sloths including the very large *Megatherium*, glyptodonts, Irish elk, cave bears, North American camels and even the giant moa bird of New Zealand. Another South American mammal, *Toxodon*, the size of a hippopotamus and considered by Charles Darwin to be "perhaps one of the strangest animals ever discovered", also disappeared. Many of its remains have been found accompanied by arrow heads. Neanderthal man also became extinct during this period.

In all this carnage, there was one fortunate survivor – the horse, *Equus*. The many species of the earlier Cenozoic living in temperate climates had been whittled down to only one by the time of the Pleistocene. Originally a North American animal, it hopped over a temporary bridge into Asia 2.6 million years ago. It subsequently became extinct in North America, but was reintroduced by European colonists with amazing success. (Note that zebras are also *Equus* species.) Cold-adapted survivors include the moose/elk (the largest remaining deer species), the musk ox, the American bison, the wolverine, the wolf and of course the polar bear.

These extinctions are widely attributed to the hand of man, as these groups of animals had survived many previous glaciations. Fully modern *Homo sapiens* only emerged from Africa about 100,000 years ago, so towards the end of the Ice Age, many of these now-extinct animals would have had to face him for the first time. Many of them were large (plenty of meat) and dangerous. *Megatherium* only died out 10,000 years ago, just as modern man was appearing in South America. The wooly mammoth could evidently fend off the Neanderthals, whose skeletons show much evidence of fractures due to close encounters with their prey. Evidently the mammoths could not cope with modern man, armed as he would have been with spear throwers to kill from a distance.

Meanwhile, mankind flourished, and within a few thousand years was to start a most impressive phase of construction, the remains of which – say, the Great Pyramid of Ghiza – are still with us today. It is known that many other animals – say, dolphins – are intelligent, but it was man alone who could create infrastructure on this scale. It appears that dolphins use their large brains to manage their complex sonar system, and after all they live in an environment where the construction of major infrastructure would be very difficult. Moreover they simply do not have the physical dexterity needed for the job!

The postglacial **Himalayas**. Braiding on a river indicates a steep bed.

****

In Britain, the ice did not even begin to melt until about 15,000 years ago, but once it began to recede, it did so quickly, in chaotic conditions of flood and torrent. This postglacial warming was abruptly reversed by an event known as the Younger Dryas between 12,800 and 11,500 years ago, when temperatures dropped by about 9 degrees Celsius in Britain (15 degrees Celsius according to Chris Stringer). Studies of ice cores from Greenland show that this extreme oscillation from balmy to freezing conditions in England took place in only ten years.

It used to be thought that this was caused by the sudden fracture of an ice dam holding up the massive periglacial Lake Agassiz in North America, the idea being that a huge amount of fresh water cascaded down the St Lawrence Valley, entering the Atlantic and bringing the

Gulf Stream to an abrupt halt. Recent evidence on the ground however suggests that Lake Agassiz drained away north-westwards. It seems more likely that the Younger Dryas was just another lurch in the climate, many of which have now been recorded within the ice age itself. It may have been caused by a movement of a mass of ice from the Hudson Bay area into the Atlantic, stopping the North Atlantic oceanic circulation, as appears to have happened on many previous occasions. Whatever the reason, worldwide temperatures suddenly fell, and took 1300 years to recover. At the end of the Younger Dryas, temperatures in Greenland rose by 7 degrees Celsius in just three years! Only a complete change in atmospheric conditions could bring about a change so quickly.

There have been many attempts to explain sudden rises in ice age temperatures such as the one occurring at the end of the Younger Dryas, but the best approach is probably once again a uniformitarian one, because there is a similar phenomenon occurring today – El Niño. This is the intermittent reversal of the atmospheric and oceanic circulation in the Pacific Ocean. During normal conditions, there is low pressure along the equator over Indonesia and the western Pacific. The surface winds are the Trade Winds, easterlies and south-easterlies which converge at the equator – the "Inter-tropical convergence zone" (ITCZ), causing wet weather in the western Pacific. Convergence is a meteorological term meaning the build-up of air either by the confluence of two currents, or the compression of one into a smaller space, and is normally associated with low pressure and rain (divergence is the opposite). (The ITCZ used to be called the Inter-tropical front, but only in some places is it a front, as in West Africa where the hot dry Tropical Continental air from the Sahara meets the warm wet Tropical Maritime air from the Gulf of Guinea. In the Pacific this is simply the confluence of two tropical maritime currents.)

The easterly winds create the easterly Equatorial ocean current. This creates upwelling off the coasts of Peru and Ecuador and brings nutrient-rich cold water to the surface, flowing up the coast of Peru as the Humboldt Current and increasing fishing stocks. The western side of the equatorial Pacific is characterized by warm, wet low pressure weather as the collected moisture is dumped in the form of typhoons and thunderstorms. The ocean is some 60 centimetres (24 in) higher in the western Pacific as the result of this motion.

During El Niño, this pattern reverses. A high atmospheric pressure cell forms over Indonesia and the western Pacific; air pressure falls over Tahiti and the central and eastern Pacific. As a result, the Trade Winds weaken or change direction and become westerlies. This westerly flow hits Peru and causes rain in the coastal deserts, while the western Pacific area misses its normal rainfall and experiences drought. The ocean currents follow suit; the Equatorial Current reverses and carries nutrient-poor tropical water along the equator, suppressing the upwelling of the Humboldt Current. The reasons for El Niño events are not really understood, but the symptoms and effects are clear. There has been a reversal of the normal circulation.

Something similar must have happened at the end of the Younger Dryas, operating in a north-south orientation rather than an east-west one. Climatologists recognise just four kinds of air: tropical continental (warm, dry), tropical maritime (warm, wet), polar continental (cold, dry) and polar maritime (cold, wet). The mid-latitude Westerlies are where the tropical air meets the polar air. Great eddies are formed at the interface, bringing the traditional wet weather to western Europe. In the middle of all this runs a very fast, narrow stream of air, the jet stream. This can shift its position by hundreds of miles to the north or south between years even today, dragging the low pressure cells and the rainfall with it. At the end of the Younger Dryas, driven by fierce heat in the tropics, it must have made a great jump northwards, bringing tropical maritime air with it into the northern latitudes, pushing back the polar air and driving the Gulf Stream before it. This would represent a reversal of the circulation where the polar air moved south.

In any event, by 11,500 years ago all the permanent ice had gone, but continuous habitation of these islands had already begun. The evidence for it includes some English cave art, with an image of an ibex or deer, at Creswell Crags in Derbyshire, dated to 12,000 years ago. Human bones of around the same date have been found at Gough's Cave, in Cheddar Gorge.

There is a theory which proposes that the Black Sea basin was flooded by water from the Mediterranean about 7600 years ago as the ice melted and sea levels rose – thereby creating the Biblical Flood of Noah. However this is disputed. As there is substantial river drainage into the Black Sea, it is quite likely that a river exit through the Bosphorus already existed during glaciated times. As sea levels rose this could have been drowned to create the straits which exist there now, turning the Euxine Lake into the Black Sea.

In more recent times a slightly warmer period lasted from 950-1250 AD, known as the Medieval Warm Period. This was followed by the Little Ice Age, from the 13th or 14th century to the mid 19th century, which was a period of significant cooling. The coldest phase of this was the Maunder Minimum, 1645-1715, thought to be related to a shortage of sunspots.

After the ice sheets had gone, much of the British land surface was left covered in the boulder clay (till) left behind by the melted ice. The extent of this is not to be underestimated. Approximately one-eleventh of Britain would be under water without it, and a further one-sixth of the country has a cover so thick that it obscures the underlying surface features. One of the best examples of a till-plain is afforded by the low plateaus of East Anglia. The most widespread deposit here is the "chalky boulder clay", an unstratified stony clay containing many flint and chalk fragments, and larvikite rocks from distant Norway. It commonly reaches thicknesses of 30-45 metres (100-150 feet), and more in higher areas as around Bury St Edmunds. Dating from the final glaciation, it overlies an earlier and deeply weathered deposit, also boulder clay, known as the Norwich Brickearth. This is notable for its content of Scandinavian erratics. The highest land in East Anglia, the Cromer Ridge, which is 90 m/282 feet high and 14 km/9 miles long, is in fact a moraine. Boulder clay is also widespread in Leicestershire, Yorkshire, Lancashire and Cheshire.

The ice sheets left other more localised deposits behind, notably those laid down in periglacial conditions. These include several levels of gravelly terraces along the Thames. In Yorkshire, the Vale of Pickering was the site of a glacier-dammed lake, and so is extremely flat. Erratic blocks – lumps of rock taken away from their original outcrops and dumped elsewhere by the ice sheets – scatter the northern parts of the country. Very noticeable are those found on the Great Limestone around Ingleborough in north Yorkshire. These are mostly Yoredale and Millstone grits, but the protection they offer means that they now stand on pedestals about 30 cm (a foot) tall, meaning that 30 cm of limestone has been dissolved around them in the 12,000 years since the end of the last ice age.

The country to which the people returned did not look quite the way it does today. In general, the south-east sank into the sea and the north rose out of it by isostatic recovery from the weight of the ice. The coastline took on an approximation to the one we see today, but with important differences. There has been much erosion in east Yorkshire and East Anglia since those times and indeed the important East Anglian medieval town of Dunwich now lies on the sea bed. The Humber Spit and the Romney Marshes in Kent did not exist. The fenlands inland from the Wash were totally undrained and in practice stretched right up to where Stamford and Huntingdon now stand. The extensive marshes in Somerset were also unsuitable for much habitation at this time, and large parts of Staffordshire and Lancashire were unusable bogs. The Norfolk Broads were simply peat beds which were dug out to create the lakes found there today.

In other countries the melting ice also had a dramatic effect on the landscape. In western Montana, in the United States, a periglacial lake formed in the middle reaches of the drainage of the Columbia River and its tributaries, known as Lake Missoula. The periodic rupturing of the ice dam which held back the lake resulted in a series of catastrophic floods downriver, creating the scablands of the east of Washington state, a deeply dissected landscape cut largely in basalt flows and unconsolidated loess. This is a landscape which owes its morphology to catastrophism (very much on the lines of the Noachian floods as envisaged by Buckland) rather than the gradualism of Charles Lyell.

North America also has a vast covering of glacial till, extending southwards into states at the latitude of Indiana. Intriguingly, the till found here has been a surprising source of recoverable diamonds and gold, brought down from source rocks in Canada. All diamonds originate in kimberlite diamond pipes, shot up from deep in the earth, but it the case of the Indiana diamonds, this has never been located. It must be buried under glacial debris, somewhere in Canada! The till as a whole represents the removal of large amount of Canadian soil and rock to the United States, and there is no doubt which country is the beneficiary. Boulder clay represents freshly ground rock exposing minerals to weathering for the first time. It was deposited over Palaeozoic strata in the American Midwest and on the whole gave it greatly increased fertility.

****

The vexed and controversial question of global warming crosses boundaries with the above discussion of the many issues involved in ice age geology and climate. It is perfectly clear that we are now only in an extended interglacial and that the fundamentals as laid down by the configuration of the continents and the Milankovitch cycles mean that at some point the earth will revert to glacial conditions. That point may be 10,000 years away for all anyone knows. However the global warming people cannot wait that long and threaten a climatic catastrophe over a much shorter time period, to be measured in hundreds of years at the most. Nevertheless, the earth at the time of the Paleocene-Eocene temperature maximum of 28 degrees Celsius (compare 14 degrees

Celsius today) was not, by all accounts, a bad place to live. On the contrary, magnolias grew on Greenland. Gaia did not close down. Furthermore, with every degree of warming comes an average one percent increase in rainfall. So why worry?

The problem once again may be the configuration of the continents. At the start of the Eocene, Pangaea had long gone, and open ocean was flowing round the continents. The Antarctic was not yet positioned over the south pole. The disposition of the continents over the face of the globe today is actually much closer to Pangaea, as sixty percent of them are joined together in the World Island. That situation recalls the Permian – and we really do not want to go back to "Permian conditions plus ice sheets".

****

Louis Agassiz was of course the man who first conceived of the ice ages, and I once visited his territory myself, in 1964 when I was only fifteen, in the Alps of Valais in Switzerland – I was one of a school party. One day we took a trip to a local glacier. The moraines, the deposits dumped by the glacier as it had retreated, were strewn over the valley. I found the glacier itself – in the summertime – rather a shabby-looking thing, with transparent, half-melted ice at the snout, covered in debris. Wearing solid but normal shoes, I clambered on a ledge, eight metres (25 feet) above the glacier. A stream of meltwater was running across the ledge, and I would have to cross this to continue. As soon as I had both feet in it, I started to slide to the edge. I was terrified – there was nothing I could do to stop myself going over the edge. Suddenly the sliding stopped and I was able to get away. But that would have been a very nasty fall, a broken ankle at least, if not a broken neck. And it seemed as if I had been stopped by a magic brake.

Chapter 18 – Erosion and Landforms: Humid, Periglacial and Glacial

Having concluded the above review of stratigraphy, it is time to take a look at geomorphology, the scientific study of landforms. A suitable place to start is the point where the geology has done its work and created a structure. What happens next? The answer is weathering.

Geomorphologists make a distinction between weathering and erosion. Weathering is the decay of rock *in situ*. The result of weathering is the accumulation of a mantle of waste or regolith. Weathering is the first part of erosion, which then involves the movement of the waste so produced. The next stage is deposition. The overall effect of erosion is described as denudation, the lowering of the surface of the land, and ultimately the movement of it into the sea. Weathering is of three types, mechanical, chemical and biological.

Mechanical weathering includes wind abrasion, especially in deserts; freeze-thaw which is so powerful it can split granite, and leads to the formation of impressive scree slopes as seen descending into Wastwater; abrasion as one rock smashes against another in a river bed; and abrasion again as rocks embedded in glaciers and ice sheets gouge out bedrock beneath them. A further form is battering by waves at the coast, leading to marine erosion.

Chemical erosion includes hydration, a reaction when water is added to exposed rock. The best-known example is the conversion of calcium sulphate from its unhydrated form (anhydrite) to its hydrated version (gypsum). Again, the iron ore haematite (Fe_2O_3) hydrates to limonite. Hydration frequently acts upon feldspars and micas. A second form of chemical weathering is hydrolysis, a chemical reaction of water on raw minerals. For example, this produces potassium hydroxide, which is then subject to further reactions to form soluble potassium carbonate. This is the type of reaction which rots the feldspars within granites to residual clay minerals, notably kaolinite. A third form of chemical weathering is oxidation, which particularly affects iron compounds (such as iron pyrites, FeS_2) when they are exposed above the water table, and results in a change in colour from blue, grey or green into red or brown. "Vadose" water, between the ground surface and the water table, assists in this process. In tropical soils the oxidation of iron can lead to the formation of iron pans or laterite. Fourthly is carbonation, a process involving the dissolution of rocks containing calcium carbonate by rain containing dissolved carbon dioxide. This reaction produces calcium bicarbonate which is soluble in water and so creates holes in the rock. Limestone, chalk and dolomite are particularly affected. The final type of chemical weathering is solution, the removal of soluble material from rocks and soil by percolating water, a process known as leaching.

Minerals vary greatly in their resistance to chemical weathering. Quartz, muscovite (white mica) and orthoclase feldspar are much more stable than olivine, augite and plagioclase feldspar. It is noticeable that the first group are light-coloured whilst olivine and augite are dark, and in general light minerals are more stable than dark. However everything depends on the circumstances. In the abundant moisture and high temperature of the tropics, a light-coloured but well-jointed granite will weather, whereas in a temperate climate, a dark igneous rock such as the gabbro of the Black Cuillins of Skye is very resistant to chemical weathering.

Biological weathering includes the loosening of bedrock by plant roots; the activity of earthworms in breaking down the soil; the work of bacteria, algae and mosses which can break down rock silicates directly; and the decay of dead plants and animals which produce chemicals – ammonia, carbon dioxide, nitric acid and humic acid – which then promote chemical weathering.

A polished, decorative black **gabbro** – it certainly looks resistant to weathering

Climate exerts a great influence on weathering and erosion. Frost weathering only occurs when there are freeze-thaw cycles; insolation weathering, shattering from the direct heat of the sun, requires very hot days and cold nights; sandblasting in deserts obviously requires desert conditions; and all types of chemical weathering work much faster in warm climates, because the speed of the chemical reactions roughly doubles for every 10 degrees C rise in temperature. Wet climates are also required for chemical weathering. In temperate climates, all types of weathering are found, but chemical weathering is more important than mechanical, as it assisted by the soil and vegetation cover, allowing the infiltration of water and humic acids into the bedrock. Such is the influence of the climate upon erosion that climatic types of erosion can be defined, each with characteristic assemblies of landforms. However there have been such dramatic changes in climatic conditions during the Quaternary that many modern landscapes are known to have evolved under a previous climatic regime. The landscape as visible in England, for example, is only 10,000 years old.

Once weathered out, the regolith is then subject to mass-wasting or mass movement, the inevitable downhill movement of material *en masse*. In temperate climates the most important form of this is soil creep, which can take place at imperceptibly slow rates – raindrop by raindrop – on the gentlest of slopes. More dramatic forms include mudflows, slumps and landslides. Slumps are especially important on cliffs and escarpments. As it moves, the earth can be rotated upwards.

****

Humid tropical climates promote active chemical decay, the products of weathering often remaining *in situ*. This can lead to the decomposition of hard rocks such as granite, forming soils 30 metres (100 feet) or more deep. Fluvial erosion and transport is the dominant geomorphological process. The soils themselves are easily leached of soluble silica, leaving behind widespread laterite formations. In wetter areas, however, the whole system is rather stable from a geomorphological point of view. The products of weathering often do not move far, forming deep soils and providing nutrients for the dense vegetation which eventually returns them to the soil. Laterite becomes indurated when the forest cover is removed, either by climate change or by the hand of man, and it is a great menace to agriculture.

The characteristic landform of drier parts of the tropics is the inselberg. This is an isolated hill, knob, ridge, or small mountain that rises abruptly from a gently sloping or virtually level surrounding plain. The term "monadnock" is usually used in the United States, where there is a Mount Monadnock in New England. Inselbergs can also form it temperate climates – Ingleborough and Pen-y-Ghent in the Yorkshire Dales are inselbergs, having caps of Millstone Grit lying on top of Carboniferous limestone.

In southern and southern-central Africa inselbergs are also known as kopjes or bornhardts. These are domes composed of granite or gneiss, rising in spectacular fashion in curved summits above near-level surrounding plains. They are thought to be formed by a process of exfoliation. The rocks become jointed in lines parallel to the surface as the overburden of later rocks is removed. The top layer of rock peels off as weathering works at the joints. In some cases it appears that the weathering has taken place underground in more humid conditions than exist now, and has been followed by erosion to reveal whaleback-type inselbergs underneath. There are examples of this in the Tsavo National Park in Kenya. As the African inselbergs sit in landscapes which are millions of years old – going back to the Cretaceous without marine incursion – it is possible that many of them formed in this way.

I have been to Tsavo myself. I can't say I noticed the inselbergs at the time, but I could hardly miss the red elephants. I have seen many African elephants, but only in Tsavo were they red. Apparently the coloration comes from the muddy water in the wallows.

Bornhardts are common structures in sub-Saharan Africa, often in areas of acacia bush where their appearance may reflect an older climates. However easily the most famous inselberg in the world is Ayers Rock, or Uluru, in the Northern Territory of Australia. This is composed not of granite but of arkose, a feldspar-rich sandstone.

****

In 1973 I arrived in Zambia in Africa to begin my first real job as a Land Use Planning Officer in the Department of Africa. I was just 23 years old, but I did know something about the job. One of the subjects I had studied for my Master of Science degree was tropical soils, and this part of agriculture would be my responsibility in the Eastern Province of Zambia. I was soon rather surprised to find that the local classification of soils looked much as it would in England. There were sands, loams and clays, and mixtures between them, such as sandy clay loam. A number was attached, 1,2 or 3, to indicated quality rather than texture. For example a nice flat, deep clay would be designated C1. A stony, hilly, boggy or shallow sandy soil would be S3. The distinction in textures was important, because there were only two real commercial crops grown in the province – maize and tobacco. Maize grew well on clays and loams, but tobacco preferred a well-drained sandy soil. Precipitation on the higher, farmed areas of the province was not heavy, an average of about 62 cm (25 inches) of rain a year, so there wasn't much problem with leaching or laterite formation.

The departing Land Use Planning Officer, Paul Barnet, tried to hand over as much knowledge of the job as he could in the one month that we had together, and he was a hard, hard taskmaster. The owner of a local Indian store was trying to diversify into agriculture, and he had identified a farm about fifty miles north of Chipata where we all lived. This farm had been abandoned after the colonial era ended in 1964, and a corner of it was occupied by squatters. One of my first tasks in Zambia was to accompany Paul and the trader, called Khalid, in conducting a soil survey of this farm. The heat of the day was dreadful, it was the end of the hot season, and it was Ramadan; Paul trailed us round the elephant grass on the farm for hours and hours – I later found out that he had a reputation for this kind of thing – what had I let myself in for? Was I supposed to do this sort of thing when Paul had gone home? When we finally finished, Paul and I were able to refresh ourselves at a well, but Khalid took nothing. He was a martyr to his religion on a day like that! Rather to my surprise, as it had not been that obvious, Paul graded most of the farm as C1 - first-class land for maize. Khalid bought the farm, installed an experienced farm manager, another Indian, and somehow booted off the squatters.

However, Khalid did not prosper on the farm. Everything went fine until the first, hefty crop of maize had been harvested and bagged awaiting transport for sale. The entire crop was then set on fire. The next year, exactly the same thing happened, though the crop was guarded, and Khalid gave up the effort. It was pretty clear who was behind this – the squatters.

****

Another climatically-defined group of landforms is the periglacial, developed in the permafrost areas adjacent to ice sheets. Two climatic extremes have been identified, maritime Arctic (Icelandic) and continental Arctic (Siberian). In the first, precipitation is heavy and mainly takes the form of winter snow, and temperatures are comparatively mild – they may average above zero over the year. At the other (Siberian) extreme, precipitation is light and falls as summer rain. Winter temperatures are severe and can reach -60 C, but summers are relatively mild and always above freezing in July. In both cases, frost action produces spreads of blocky and finer debris from the disintegration of larger boulders. The process of "cryoturbation" (by frost freezing) churns the top levels of the ground in the melting season. Chemical weathering is at a minimum as reactions work slowly in the cold.

Periglacial climates are distinguished by the presence of permafrost. The important point about this is that there is a summer thaw, causing the top layer of the soil to shift, sometimes in distinctive patterns. In spring and autumn there is a diurnal freeze-thaw cycle. This accelerates mechanical weathering. There is also free water at these times, and the water acts as a lubricant, causing soil actually to flow downhill, a process known as solifluxion. Nevertheless the base horizons (levels) of the soil remains frozen throughout the year so overall drainage is poor and much of the ground is swampy or waterlogged in the summer.

Frost heave and frost thrust move rocks around in the soil mixture, which churns and can form distinctive dog's tooth patterns ("involutions") in vertical section. The ground surface becomes highly irregular or hummocky. Large flat-floored oval depressions can form, with a diameter of 15 kilometres (10 miles) and a depth of 40 metres (130 feet). In addition to this vertical sorting of the soil and rock material, lateral patterns also form, known as stone polygons. These are hexagons or near-circles of stone fragments with a central area of moist mud. On steeper slopes they elongate into stone stripes.

Solifluxion takes place when the whole body of the soil moves downhill, often on quite modest slopes. This movement can achieve rates of 2-5 cm (1-2 inches) a year on slopes of 10-15 degrees. This contrasts with a movement of about 2 mm (less than an eighth of an inch) a year for the soil creep which takes place in temperate climates, and so can be twenty times as fast. The accumulated material of solifluxion gathers in valley bottoms. An unstratified debris of chalk and flint fragments has formed from this cause in the bottom of many Chalk valleys in the south of England, known as Coombe Rock; it can be several metres thick. Also large blocks of stone can be exhumed at the surface and moved downhill. Such are the sarsen stones of the Marlborough Downs in Wiltshire, which were used in the construction of Stonehenge. This rock, indurated by a silica cement, is a form of silcrete (silicon concrete). It formed sheets in the Palaeocene which were broken up in the tundra conditions of the Ice Age. Locally the sarsen stones, which are found scattered all over the Chalk Downs, are known as greywethers because from a distance they look like sheep.

Some features of periglacial conditions have been widely observed but defy an obvious explanation. One of these is the asymmetrical valley, where one side has much steeper slopes than the other. This can be observed in the Chilterns, where the valleys run from NW to SE. The slopes facing SW – that is, the sun and the wind – are much steeper than those facing NE. Another process which is not well understood is the formation of nivation hollows, quite deep depressions generally on north-facing slopes occupied by snow for most or all of the year.

Wind erosion in periglacial conditions produces loess deposits, very widespread in places. Loess is a brownish-yellow sandy loam, rich in lime and homogeneous in structure. Across Europe it occurs in a belt running from northern France (where it is known as *limon*) into Belgium, Germany and Poland. Usually it is 3-6 m (10-20 feet) in thickness. Far deeper deposits occur in China, where the loess soils form valuable farming land. Locally they can reach 300 metres (1000 feet) in depth, smothering the pre-existing landscape. In China loess formation is not entirely a periglacial phenomenon, however, as it still takes place when the dusty winds blow out of central Asia and the Gobi desert, across to the northern reaches of the Yellow River and to Peking itself. Loess deposits are also found in the Mississippi Valley where they are called adobe.

****

No form of erosion has such a dramatic effect on the landscape as glaciation. Direct action by rocks embedded in the ice wears away U-shaped valleys, corries and even the basins of major lakes such as Lake Geneva. Where great ice sheets have melted and the process of subsequent landforming has been slow, a vast wilderness of thin soils, morainal dumps, thousands of lakes and a badly disorganized drainage network is left behind. One look at a map of Finland or Canada is enough to show the chaos that has been served upon the land. In Finland, which after all is hardly a small country and has water everywhere, it is impossible to discern a river of any length. The pre-existing relief has been streamlined and bulldozed. Parts of Scotland are similar. In Lewis, the famous Lewisian gneiss has been abraded into numerous rocky hills ("knocks") intermingled with hollows containing small lakes ("lochans"). The soils are patchy or non-existent as soil-forming processes act only slowly in this bleak area. In other glaciated areas, there are numerous rocky cliffs which have been produced by the plucking away or "quarrying" of large joint-bound blocks.

Glaciated valley at **Hallstatt, Austria**

Soft snow compacts into a granular mass called firn or *neve*, and gradually becomes frozen solid to form glaciers. These have terrific erosive power and have left behind a characteristic assemblage of landforms, best seen in mountainous areas. These include the U-shaped valley, gouged out deep by a glacier and subsequently filled in flat at the bottom by lacustrine deposits. These are nowhere better seen than along the coast of Norway or Alaska, where rocky fjords descend thousands of feet from the coastal mountain tops to far below sea level. The characteristic U-shaped valley not only has steep sides but also will have been straightened out, exhibiting truncated spurs. These valleys also frequently display headward steepening. The glacier deepens its bed to an extent which could never be achieved by a mountain stream. A glacier bed may also typically feature rock steps, which the glacier itself tumbles over in a frozen, crevassed torrent. These are normally associated with resistant bands of rock.

An especially fine and indeed famous example of a U-shaped valley is found at Yosemite in eastern California. The national park is only small, in fact only one valley, but the landforms are staggering. One of them is the Half Dome, a granite dome which has literally been sliced in two. Part of the same phenomenon is the hanging valley, occupied by a stream left far above the valley floor by the gouging. Higher up are found arêtes, sharp ridges along the tops of hills, and corries (cirques, cwms), scooped-out hollows at the tops of mountains, often now containing a tarn. Pyramidal peaks such as the Eiger or the Matterhorn are a consequence of corries and arêtes. *Roches moutonnees* are boulders which are smooth on the upstream side, but plucked out on the lee side. Then there are kettle holes, rounded holes left behind where a lump of ice melted in the ground. Individual outcrops of rock in glaciated areas often show signs of striations, where other rocks embedded in the ice have scratched them.

Other features include drumlins, parallel, whale-shaped mounds perhaps 10 metres (32 feet) tall and 200 metres (650 feet) long, resembling, according to some, a basket of eggs when viewed from above. They result from a rippling effect on softer deposits beneath the ice. Scots may recognise these, as part of the city of Glasgow is built right on top of a "swarm".

The Half Dome, Yosemite

Beyond the mountains, glaciers can spill out onto the plains to form piedmont glaciers. One of the best-known of these is the Malaspina Glacier of Alaska. Then regional or even continental ice sheets can form. In these the movement of the ice is radial as the ice moves from points of high to low pressure within the mass. The surface can remain almost stationary whilst the underside of the ice moves, very slowly. Piedmont glaciers and ice sheets can gouge out large lakes beyond the mountains. These include the Great Lakes of North America – surely the most impressive of all Ice Age geological achievements – and Lakes Geneva and Garda. Lakes of this type are considered by geomorphologists to be essentially temporary features of the landscape which will fill in eventually, but in the case of Lake Superior, that will take time!

As well as rock carving there are also deposits. Moraines are rock and clay debris or till accumulated by the glaciers, and then left behind.

These come in lateral, medial and terminal forms. Cape Cod on the north-east coast of the United States is actually a medial moraine.

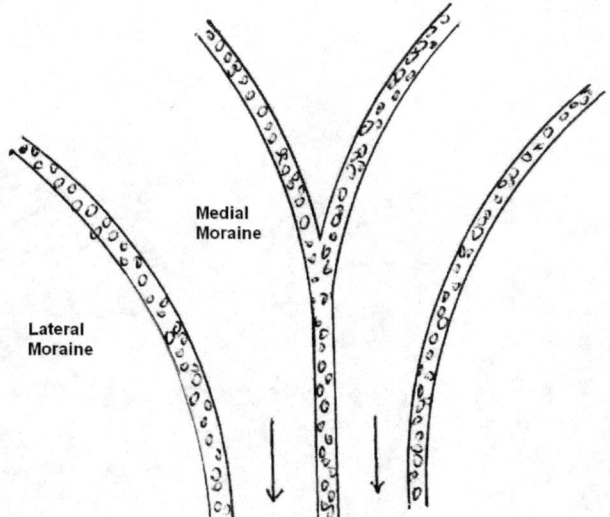

Medial moraines form where two glaciers merge together

Discharging rivers around glaciers leave landforms of their own, stratified deposits known as kames (hummocks) and kame terraces (formed at the sides of glaciers) as well as outwash fans, some of them of very great extent. The outwash fans from the ice caps of Iceland are a truly impressive spectacle and form a fair proportion of the country's farmland. A feature similar to a kame is an esker, a sinuous ridge of sand and gravel formed at the margins or even underneath stagnant ice. Its linguistic root is an Irish word for a path, as the eskers made good pathways through the bogs.

There is also much evidence of temporary lakes ponded up against the ice. One of the best-known is at Glen Roy in Scotland, where the Parallel Roads are a well-preserved set of strand lines from the lake.

Another glacial landform is *crag and tail*, where an ice sheet selectively erodes softer sedimentary rock around an upstanding resistant rock, then leaving a tail of protected softer sediment behind the prominence. A good example of this is Edinburgh Castle, which stands on a volcanic plug – the crag. Behind it lies the tail, on which the Royal Mile is built, running down eastwards to Holyrood Palace.

Boulder clay country does not necessarily make good farmland, particularly when of the "kettle and kame" variety, hummocky with sinuous ridges and boggy lakes strewn with erratic boulders. Around the city of New York a surprising amount of this type of country has been reserved for cemeteries!

Erratic boulders are a very common feature of glaciated landscapes and have been mentioned at various points in the book. They are of a completely different lithology from the surrounding country rock, as with the Scandinavian granites dumped on the limestones of the Yorkshire Dales by the ice sheets. One slightly different form is the dropstone, where an erratic boulder has been dropped into a lake or into the sea onto sediments of a completely different type. One hallmark of the dropstone is the curvature of the sediments underneath it, pushed down by the weight of the boulder. Dropstones are a key feature of the Namibian evidence for Snowball Earth identified by Paul Hoffman. If the dropstone falls into a lake, it may distort the *varves* underneath it. Varves are laminated deposits, coarse sand when laid down in the meltwater season, then fine clay deposited as the lake ices over in winter.

Chapter 19 – Limestone, Coastal and Desert Landforms

Carboniferous limestone is noted for the vertical joints it contains. These are weaknesses in the rock, which are exploited by the agents of erosion, so leading to the most important characteristic of Carboniferous limestone – its permeability. Rainfall is slightly acidic because of the dissolved carbon dioxide it contains, and so erodes the alkaline limestone as water seeps through the joints. This creates a landscape that lacks surface drainage but which has all manner of characteristic surface and subsurface features. This is known as karst landscape, after a part of Croatia where it is widely developed.

Small surface depressions called sinkholes, which are 1-3m (3-10 feet) deep and 3-5m (10-17 feet) across, form as a result of the subsurface collapse of limestone. They are very common throughout the Yorkshire Dales. Larger depressions are called dolines. Streams flowing from higher impermeable slopes disappear into the ground through sinkholes when they reach permeable limestone. The sinkholes can enlarge into full-scale potholes, of which Gaping Gill and Alum Pot in the Yorkshire Dales are well-known examples. An extention of this in limestone country is a gorge formed as the result of the collapse of a roof above a cave system. Gordale Scar is an excellent example.

The most common examples of subsurface features in a limestone landscape are the caves. In the Yorkshire Dales these are numerous, for example Ingleborough Cave. Other noted caves are found in the Cheddar Gorge and at Castleton in Derbyshire. The caves themselves and their associated formations vary greatly in size, but they all depend on the dissolving of limestone by acidic rain for their creation. This however is a reversible process: the calcite can be redeposited as stalagmites (up), stalactites (down) and flowstone. The rate of growth of these formations is about 1cm per 200 years; it is therefore an extremely slow process. Some stalactites may be millions of years old.

Dry valleys are valleys without streams. For example a stream flows from Mallam Tarn which then disappears underground, only to reappear a few miles to the south under the massive Mallam Cove, a cliff which was one a huge ice-age waterfall. The water flowed through the dry valley in subglacial conditions when the substrata were frozen solid, and so impermeable.

A limestone pavement is an area of almost bare, flat rock developed after the surface has been exposed by the scouring action of an ice sheet. Existing joints are subsequently exploited by the action of chemical weathering to form deep slits called grykes and rounded blocks called clints. In fact limestone pavements containing clints and grykes are the most difficult imaginable walking country where a broken ankle is always threatened!

Clints and Grykes above Malham Cove, Yorkshire

Different types of karst scenery have developed around the world where limestone outcrops. One of the most fantastic formations is at Guilin in southern China, where hundreds of hills rise up sharply from the landscape like so many giant's teeth. Such hills form the background to many a Chinese painting, though it is doubtful whether the artists ever saw them. The formation here is an old, hard Devonian limestone where the warm, monsoonal climate and the absence of any glaciation has allowed these exaggerated landforms to develop.

Limestone scenery along the **Li River, Guilin, China**

Limestone scenery can certainly be spectacular but it is not necessarily much fun. Have you ever been down a pothole, for example? I went into one once, near Kettlewell in the Yorkshire Dales.

The experience involved crawling along a passage no more than a foot deep with a stream running along the bottom of it, finally to reach a cave with some stalactites and stalagmites in it. It was very wet, highly claustrophobic, and I never sought to repeat the exercise. However some people go so far as to explore the caves by diving. The chances of getting trapped if a cave fills up with water after a sudden storm make this the riskiest of sports. I know someone who did go in for this sort of thing, a man called Dave.

"Why did you give up, Dave?" I once asked him.

"All the people I used to go with were dead!" was his response. I later asked a colleague about this "sport", another Dave in fact. "That," he said, "is for nutters!"

Sea arch, **south-western Iceland**

Coasts have their own unique set of landforms, regardless of climatic zone, and they are notable for the rapidity with which they can change. Coastlines are frequently dramatic and dynamic, and are almost invariably immature, as present-day sea levels are only about 6,000 years old. For example, beaches are in a state of constant evolution, as are spits, sand bars and salt marshes. Cliffs experience sudden slumps. These features depend on the interplay of wind and wave action, tides, currents, the supply of beach material, rock type, aspect and so on. The strength of the wind and the tides is partly dependent on the fetch, or the distance over water to the nearest opposing coast. The ferocity of wave attack in south-west of England is due to the long fetch, right across the Atlantic. Paradoxically the areas experiencing the most rapid erosion are in the east, in Yorkshire and East Anglia, where the rocks are weaker. Very rapid erosion at rates of 1.5–2 metres (5-7 feet) a year over thousands of years has caused the coast of the Holderness area of east Yorkshire to disappear into the sea. This is partly balanced by the deposition of sand bars, spits, mud flats and marshes further down the coast, which is little compensation when good land has been replaced by useless tidal features.

The overreaching role of geological structure can be illustrated along the coast of Wales. The NE-SW trend of the Lleyn Peninsula in the north is determined by Caledonian folding in old, resistant rocks. Southwards the broad sweep of Cardigan Bay lies parallel to the axis of the Teifi anticlinorium (series of anticlines on a regional scale), lying 10-15 km (6-9 miles) inland. In the south the west-east trend of the Pembrokeshire peninsula reflects the continuation of the Hercynian folding of the South Wales coalfield to the east. Southwards again, the Bristol Channel itself is an exhumed structural and erosional depression dating from Hercynian and early Mesozoic times.

The coastline of Britain has been greatly affected by the dramatic changes in sea level of Quaternary times. In southern England, the Solent which separates the Isle of Wight from the mainland overlies a drowned river which used to flow from west to east. Plymouth in Devon and Milford Haven in Pembrokeshire both stand on rias, drowned valleys. In Scotland the many sea-lochs are submerged glaciated valleys, the products of fluvial and then glacial erosion rather than coastal erosion. In general, however, Scotland has emerged from the sea rather than being drowned by it. The distance between the Firths of Forth and Clyde is thought to have been less than half the modern distance only 6000 years ago; the bounce is due to isostatic recovery following the melting of the ice cap.

Beaches depend for their maintenance and growth on a supply of new material, which ultimately must come from the erosion of the coast. Sheer wave power alone is enough to take material from the unconsolidated boulder clays of East Anglia and Yorkshire. In harder formations, the waves exploit joints and faults. Sometimes this action can give rise to distinctive features known as geos, narrow inlets of the sea in hard rocks. Limestone coasts frequently feature caves, partially formed by the carbonation process of underground water, then exposed by the recession of a cliff and later attacked by the waves. Harder rocks including limestone and also chalk can be undermined directly by wave action to give characteristically vertical cliffs. The form of a cliff depends on the dip of the rocks. Horizontal strata and those dipping inland give vertical cliffs, whereas rocks dipping out to sea do not.

Cliff lines are clearly forced back by sustained marine erosion and there have been attempts to attribute planed-off bedrock platforms many miles wide to marine erosion, but it is not thought to be effective over more than two kilometres (1.2 miles) unless accompanied by a slow, steady marine transgression.

Longshore drift is responsible for the great shingle foreland at Dungeness in Kent. This is essentially a modern creation, not thought to have been there in Roman times. It is composed almost entirely of flint pebbles washed along the coast. The local rocks do not contain flint. The magnificent sandy beaches of South Wales are also due to drifting from the Severn estuary, as in some areas the local rock is limestone.

The pebbles found on any beach are likely to have a wide distribution of original sources, as these represent the hard material which has survived erosion and transportation. On the coast of North Wales for example it is quite easy to pick up a black basalt pebble (from Northern Ireland), a pink granite one (from the Isle of Arran), a brown flint (from the sea bed) and yet another white one of local Carboniferous limestone.

The most remarkable shingle bar in Britain is Chesil Bank, situated at the eastern end of Lyme Bay and stretching to Portland Bill. This illustrates a feature of such bars, pebble sorting. The shingle increases in size towards Portland. The bar also rises in height in the same direction from 7 to 13 metres (23-42 feet) above high water line.

Spits form most notably at the mouths of rivers when a longshore current drops its material, diverting the river from its direct exit to the sea. One of these is Orford Ness, which diverts the river Alde 17 km (12 miles) to the south. Other well-known spits are found at Blakeney Point in north Norfolk, and off the mouth of the Humber.

Sand dunes are another very variable – shifting – feature of the coast. They grow and are stabilised with the assistance of vegetation, namely marram and sea-couch grass. These grasses may be planted by local authorities to stabilize shifting dunes, which can be a menace. The best-known disaster was at Culbin Sands in Moray, eastern Scotland, where dunes overwhelmed an area of rich cultivated land and turned it into a desert from 1695.

In sheltered areas of the coast – especially within estuaries, behind spits and in deep bays, deposition forms salt marshes. These are then colonised by a series of halophytic (salt-loving) plants which consolidate the deposits. One of the most remarkable of these is a plant called *Spartina townsendii*, which since 1870 has spread along the south coast of England. Indeed it has been deliberately introduced elsewhere due to its unrivalled powers of trapping mud and so winning land from the sea. Mangroves perform a similar function in tropical seas.

The creeks which form within salt marshes often form remarkable fractal patterns, geometric shapes which can be split into parts, each of which resembles a reduced-sized version of the whole. This is a property called self-similarity and was described mathematically by Benoit Mandelbrot in 1975. Mathematically it is represented by an equation which features iteration, a circling round at smaller and smaller scales. Books on chaos theory and fractals often quote natural examples found in the shape of coastlines and in such things as tidal creeks.

One marine landform feature found in warm climates is the result of creative rather than destructive biological activity – the coral reef. Individual coral polyps amass in millions to build on the remains of their dead predecessors. Each polyp secretes a calcareous shell, but corals are far from the only contributors to a reef. Other creatures – algae, foraminifera, molluscs and echinoderms – also contribute their remains. Coral reefs will only form and grow in clear waters of 20 degrees Celsius or above, with a pH of above 7 (i.e. not acidic) and are generally found between 30 degrees north and 25 degrees south of the equator. They can form fringing reefs, barrier reefs or atolls. The Great Barrier Reef of Australia is 2000 kilometers (1220 miles) long with an average width of 150 kilometers (90 miles). An atoll is a circular coral reef with no central island. The theory of their formation came originally from none other than Charles Darwin, and is still supported today. He proposed that the island – normally an extinct volcano – around which a fringing reef forms slowly subsides, allowing the coral to keep growing upwards and outwards, keeping level with the surface of the sea.

****

One place which is very much a coastal creation is Florida. The state is extremely flat and low-lying, reaching only 105 metres (345 feet) at its highest point. It has no rocks older than Eocene, and most of the southern part is simply sandbanks built onto karstic limestones over millions of years. In general size and shape, Florida bears a remarkable resemblance to the Korean peninsula, and from the naval man's point of view, the two are similar. Both jut out provocatively from continents and are bordered by straits – the Florida and Korea/Tsushima – of obvious strategic importance. Indeed the Russians and the Japanese fought a major naval battle in the Strait of Tsushima in 1905. However, the similarity in size and outline is purely coincidental, because Korea is mainly Precambrian in geology.

****

View across **Inch Bay, Dingle Peninsula, Ireland**, a coast like no other. The geology is Paleozoic, with rocks from the Silurian and Lower Carboniferous.

****

In arid and semi-arid areas, physical weathering caused by the expansion and contraction of rock surfaces produces coarse debris or sand, and there are few clayey products of chemical decomposition. This is not to rule out chemical processes altogether, as even a little water – say in the form of a heavy dew – can greatly accelerate weathering. Laboratory experiments have shown that common rocks are remarkably resistant to mechanical breakdown by temperature changes alone. Also the layers of rock detached by exfoliation are too thick to be affected by diurnal changes in temperature. Chemical weathering in deserts does attack joints, shaded places and the bases of rock masses.

There is certainly much debate about the roles of the different weathering elements in deserts. Some believe that stream floods (down wadis) and sheet floods (across pediments) are the main agents of erosion, as obviously if only a small amount of annual rainfall occurs in one great storm it can have a dramatic effect on reshaping the landscape. Sand-laden wind is a powerful sculptor but more than anything the wind seems to move around what has already been weathered away. In any event, soil-forming processes scarcely operate in desert environments and in many places there is nothing but bare rock.

The landforms take on an altogether more angular aspect. However, as there have been so many climatic changes in the Quaternary period, it is not easy to know when some features formed, and indeed there are some obviously pluvial features in the middle of the Sahara desert. It is thought that the cause of this was a shifting of the main wind belts towards the equator, bringing cyclonic rain into the Sahara.

There is certainly sand in deserts, lots of it, but it is estimated that in the biggest desert of them all, the Sahara, sand dunes or ergs only form about ten percent of the total surface. The rest are stony surfaces known as regs (desert pavements) and hammadas (stone-strewn deserts). Sandstorms are nevertheless common, swirling masses of scorching air laden with dense clouds of blistering sand, feared by all desert travellers. In the sandy areas, dunes are classified into crescentic barchans, which form transverse to the wind where the wind blows mainly in one direction; and longitudinal or "seif" dunes which form parallel to the wind, where a secondary cross-wind also blows.

The mass transport of sand by wind is a great feature of the Sahara and it can transport sand as much as 2500 kilometres (1500 miles), as we all know when our cars are occasionally covered in red dust in Britain. I once took a holiday on the desert island of Fuerteventura in the Canaries. This island features large areas of sand dunes which I was told consist of sand blown over from the Sahara desert 100 kilometres (60 miles) to the east. I thought this was a historical process, but one year we went at the New Year. The whole place was obscured by a howling sandstorm, we never saw the sun, and new sand arrived in quantity!

The most important desert landform is the combination of escarpment and pediment, with a pronounced "knick" between the two, often covered by a scree fan. The pediments can form vast plains sloping at low angles in front of the steeply-rising scarps, cutting across all lithologies, hard and soft, to form flat surfaces. Naturally there has been much speculation about how such distinctive landforms have been created. Some favour parallel scarp retreat, maintaining the same angle of scarp slope, others sheet flooding, and so on.

Another common desert landform is the deflation hollow, created by the wind. The best-known of these is the Qattara depression in the western desert of Egypt, which is 320 km (200 miles) by 160 km (100 miles) and attains a depth of 134 metres (435 feet) below sea-level. Chemical decay in unresistant rocks at the base of the hollows may accelerate weathering, making wind erosion more effective. Deflation halts at the water table, so the effect of it may be to create an oasis which can be quickly stabilised by the growth of palm trees.

Pedestal rock, **Jordan desert**

The best-known landforms of the deserts and semi-deserts of the south-western United States are mesas and buttes. A mesa (Spanish for "table") is an elevated area of land with a flat top and sides that are usually steep cliffs. It takes its name from its characteristic table-top shape. A butte is a smaller version. Both are characteristic of dissected landscapes, where the dissection may have been a pluvial process.

For anyone wanting to see buttes, there is no place in the world like the Monument Valley of northern Arizona and southern Utah. It forms a region of the Colorado Plateau, characterized by a cluster of vast sandstone buttes, the largest reaching 300 metres (1000 feet) above the valley floor. The valley lies within the range of the Navajo National Reservation, and if you hire a driver to take you round, as we did, he will be a real Navajo Indian.

Unmistakably the **Monument Valley, Arizona**

The characteristic vivid red color comes from iron oxide exposed in weathered siltstones. The blue-grey rocks in the valley are coloured by manganese oxide contained within them. The buttes are clearly stratified, with three principal layers. The lowest layer is a shale, the middle a sandstone and the top another shale capped by siltstone; all of these rocks are quite easily weathered.

The area is most famous for its use in many classic cowboy films, notably as directed by John Ford and starring John Wayne, including *Stagecoach* and *The Searchers*. There is a landmark in the park named after John Ford.

The surprising thing about Monument Valley is that it is only quite a small area, surrounded by much more ordinary countryside. When I first went to this area, it was only visible from miles away, on the horizon, but looked so amazing from that distance that we determined to go back, which a few years later, we did.

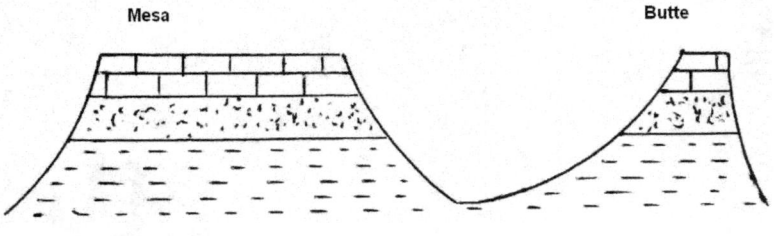

Mesas and buttes

****

Not very far away from Monument Valley is the Grand Canyon of the Colorado River in Arizona. It is a place where weathering has imposed a degree of uniformity on the modern scenery, and I went there with my wife and children in the summer of 1993 to have a look for myself. Working our way across from the geological wonders of southern Utah, Bryce Canyon and the Zion National Park, via Lake Powell and the Hoover dam, it had been a long day after an early start. We travelled in a coach with a mixed party of other foreigners, the largest group of whom were from Italy. We were supposed to rotate seats on the coach, and after we had been at the front for a few days, we offered our seats to the Italians, who had stayed in the middle. They, however, did not want to move, as this would mean breaking up the pattern they had formed. As is common with Italians, most of them spoke only Italian, and they wanted to stick together. So the coach drove on, mile after mile, ever rising, through poverty-stricken Indian reservations on poor land, with no sign whatever of the canyon, until suddenly it lay before us – the most stupendous geological sight on the planet. It was utterly breathtaking.

The Grand Canyon is 446 kilometres (277 miles) long, and with a maximum width of only 29 kilometres (18 miles), attains a depth of well over 1500 metres (one mile). It is asymmetrical in that the northern rim of the canyon stands 300 metres (1,000 feet) higher than the southern. There are only ten main rock formations exposed in the Grand Canyon itself, and the breaks between them are clearly visible. They age from the Vishnu schist, Precambrian basement rocks 1.7 billion years old, to sedimentary limestone 260 million years old at the top. Directly above the Vishnu formation there is a great unconformity, as the next rocks in the sequence are Cambrian in age. From this point upwards, they demonstrate layer-cake stratigraphy with one facies lying flat below the next. Both marine and terrestrial sediments are found, including fossilised sand dunes from a former desert. The characteristic red colour is once again simply iron staining. When iron-rich minerals weather in the presence of abundant oxygen the result is reddish ferric iron. Such red rocks often indicate deposition under terrestrial conditions.

The Grand Canyon was first explored by a one-armed Civil War veteran called John Wesley Powell and his team in 1869. They ran the 446 kilometres (277 miles) of the Colorado River, effectively white-water rafting with no aids to buoyancy. Three of the party never returned – two got out part-way and were never seen again. Powell went on to found the US Geological Survey.

The remarkable thing is that the Colorado flows through the high Colorado Plateau, starting from the eastern side, without giving up and turning back to the east. It does so because it was able to keep pace with the uplift which has affected the region from about 75 million years ago, starting in the Laramide orogeny. This mountain-building event is largely responsible for creating the Rocky Mountains to the east. In total the Colorado Plateau was uplifted an estimated 3,000 metres (2 miles).

The opening of the Gulf of California by tectonic shift around 6 million years ago enabled a large river to cut its way northeast from the gulf. The new river captured the older drainage to form the modern Colorado River, which in turn started to form the Grand Canyon.

The uniformity evident in the landscape is due to erosion in semi-arid conditions. This type of weathering is also found in other semi-deserts of the world, but nowhere is it developed to such a high degree.

The type of uplift observed on the Colorado Plateau is thought to be caused by movements within the mantle, and is described as epeirogenic; other regions, notably the West Siberian lowlands, demonstrate the reverse of it, sustained subsidence. There are other forms of epeirogenic uplift, for example in and around mountain areas long after plate tectonic activity has ceased. The heavy denudation rates characteristic of new mountains are balanced by uplift because crustal thickening persists.

Chapter 20 – Rivers

In temperate and tropical areas, once a rock has weathered, the job of moving it away from its place of origin then falls primarily to the drainage system. The first systematic thinker in this area was the founding father of geomorphology, the American William Morris Davis (1850-1934) and he still casts a long shadow over the subject. His most influential contribution concerned the cycle of erosion, which is a model of how rivers create landforms. In his terms everything depended on three factors – structure, process and stage. By structure, he meant the underlying geology including its relief; by process, the actions of erosion; and by stage, the extent to which the processes had modified the initial structure. For the sake of simplicity, Davis envisaged the rapid uplift of a flat surface. In the stage of youth, the rivers create V-shaped valleys. In maturity, the rivers carve wider, flatter valleys. The interfluves or areas between the rivers are also flattened by various processes of erosion. In old age the rivers meander across broad valleys amongst gently rolling hills. Finally, all that is left is a flat, level plain at the lowest elevation possible, called by Davis a peneplain – almost a plain, as a plain is actually a completely flat surface. After another uplift of mountains, rejuvenation occurs and the cycle starts again.

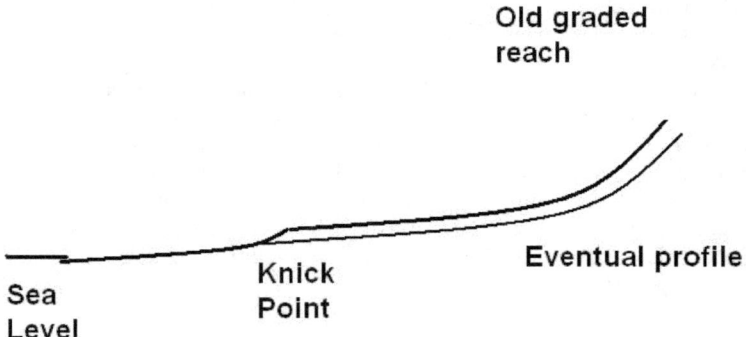

The knick point. After a fall in sea level a new profile is established for the river valley, but it has to work its way back from the sea.
****

Rejuvenations by uplift, or falls in sea level, are recorded in river valleys by changes in the slope or profile of the river bed itself. The standard grading is a concave curve, sloping steeply down at the headwaters, then flattening out in the middle and lower reaches until the base point – normally the sea – is reached. However if there is a fall in sea level, this is reflected in a deepening of the curve from the sea backwards, so that a second valley forms within the original valley at a lower level. This works its way back up the course of the river. The point at which this new level reaches the maximum point of cutting back and the river runs at its original level is called the knick point, and is often marked by a waterfall or rapids. Obviously some knick points are rather impressive, for example, Niagara Falls.

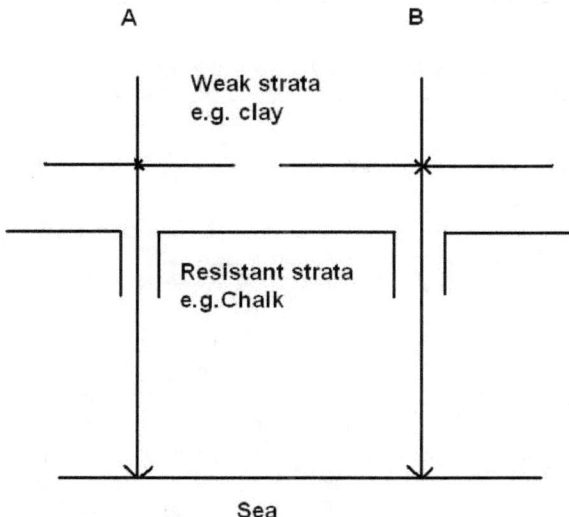

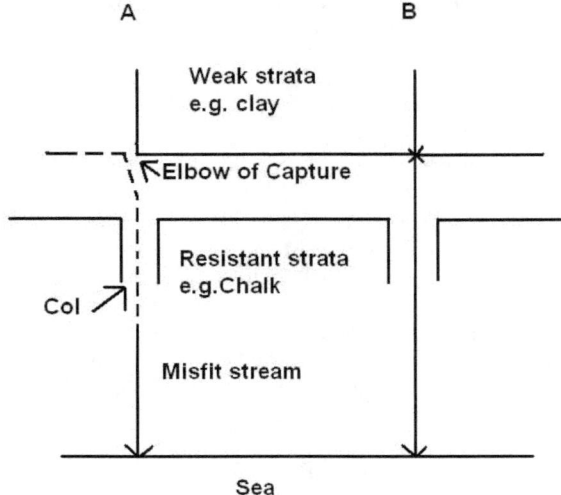

River capture. Consequents A and B drain to the sea whilst their subsequent tributaries hollow out weak rock in the clay vale. The tributary of B then captures A leaving a dry col and a misfit stream.

****

Davis also envisaged a starting structure with a series of tectonically created anticlines (ridges) and synclines (valleys), such as he found in the mountains in the north-east of the United States. Rivers would drain the synclines. However, one river would prove more vigorous than the others, having more energy (perhaps because it is larger) and cutting downwards faster. Its tributaries, running at right-angles to the main stream down the sides of the anticlines, could breach the tops of the anticlines and "capture" the similar tributaries of the next river in the system. Eventually they would capture the whole adjacent river, so that two rivers had become one – a process called river capture, or stream piracy. For this process actually to work, the pirate stream must be incised to a substantially lower level than the captured stream to give it sufficient erosional advantage. River capture is often quite clear on maps because it leaves some characteristic features behind. One is the elbow of capture where a river makes a sudden change of course, often at right–angles. This marks the point at which a stream running say south-north is captured by a tributary running west-east, and so starts flowing eastwards instead of southwards. Beyond it in its old course may be a much smaller stream or "misfit". In anticlinal country there may also be a col or gap in the anticline where the captured river used to run, now unoccupied by a river of any size. A well-known example of a col within England is the Farnham gap in Surrey, where the River Wey, which used to flow northwards through the gap, makes a right-angled turn in front of it and flows off to the east instead.

Knick point of the Rhine, the Rheinfall at Schaffhausen, Switzerland

****

Davis defined further stages beyond this. Because an anticline is structurally weaker than a syncline, having stretched rocks at the top, eventually the stage would be reached where the anticline would lie below the level of the syncline – what is known as a denuded anticline. The best example of this in Britain is the Weald, where older rocks are exposed in the centre, below the newer rim of Chalk. At this stage the main river will lie in the centre of the anticline.

A river which flows from an uplifted structure by gravity to the sea is said to be "consequent", that is, its course is a consequence of the initial structure, be that a folded anticline structure such as classically exists in the Jura Mountains, or a plain of marine erosion which cuts across lithologies of different hardness. In the case of a planed-off surface disguising numerous folds, the consequent is said to be discordant with the structure. Tributaries flowing down the sides of anticlines are said to be secondary consequents as their courses also follow from the initial structure. After the initial stage, tributaries known as "subsequents" begin to exploit the underlying geology, seeking out zones of weakness such as faults, aniticlinal axes and softer rocks such as clays and unconsolidated sands, and bringing the drainage system into closer "accordance" with the underlying lithology. This process is known as "adjustment to structure". It is certainly the case that weak anticlinal ridges can be eroded below the level of the original syncline, a feature known as "inversion of relief". A river flowing on or close to the axis of a denuded anticline is certain to be a subsequent.

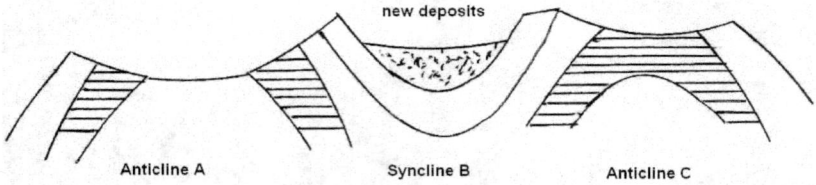

Anticline A   Syncline B   Anticline C

Inversion of relief: syncline B is now at a higher level than anticline A. A river flowing in the valley of syncline B is consequent, but a river flowing in anticline A is subsequent.

The rivers of the great plain of England exhibit a high degree of adjustment to structure. The lower Severn, the middle Trent and the upper Thames have sought out courses in weak rocks – Triassic and Lias marls and clays, the Keuper Marl and the Oxford Clay respectively.

In the course of time the subsequents may become the main rivers, capturing some of the original consequents and flowing along one of the above lines of weakness. At an even later stage the river may resume its original course at the bottom of the syncline, which at the previous stage formed the higher ground. In this case it is called "resequent". More mature drainage systems also contain streams defined as "obsequent", running in the opposite direction from the original consequent down back slopes.

If a river shows discordance with the underlying geology it is likely to be a consequent river – if, for example, it cuts through a series of denuded anticlines at right-angles then it can hardly be described as subsequent. Its course may also pre-date folding or uplift, the river maintaining direction by keeping pace with the uplift. This phenomenon is known as antecedent drainage and is exhibited by some large rivers, including the Rhine in its gorges between Mainz and Bonn. "Superimposed" drainage is similar, found where a river which developed on one lithology (say chalk) removes the cover to reveal an unconformable and entirely different structure underneath (say Coal Measures). It will the run discordantly across the Coal Measures although it ran accordantly across the chalk. This sort of thing is thought to have happened along parts of the courses of the Aire, Calder and Don Rivers in Yorkshire, described as "Chalk consequents", although the chalk has long gone. The difference between antecedent and superimposed drainage is that in the former, the rivers are older than the underlying structure, whereas in the latter, the rivers are younger.

The term *superimposition* was originally coined in 1875 by the American explorer and geologist, John Wesley Powell, to describe the drainage east of the Rocky Mountains towards the Missouri River, as exemplified in the state of Wyoming. It is believed that up to 5 million years ago, the eastern Rockies had eroded so far as to produce a relatively gentle slope down towards the east comprising a mass of sediments on top of ancient mountains. In this way, Wyoming would have resembled Iowa today, plus a few peaks. After this time uplift occurred in the eastern ranges of the Rockies (collectively known as the Foreland Ranges, including the Laramie and the Bighorn Ranges). This phase is called the Exhumation (of the ancient mountains). The pre-existing rivers including the North Platte and the Bighorn stripped away the soft, recent sediments and then maintained their courses as they cut through the much more resistant ancient mountains beneath, creating deep gorges as they did so. A perfect example of this lies on the Sweetwater River, which cuts a gorge 90 m/300 feet deep through a granite ridge, Devil's Gate, even though there is a way round less than half a mile away. However in the fork between the North and South Platte Rivers is an area called the Cheyenne Tableland. Here there are no large rivers and the ground surface slopes gently up to the Laramie peaks – it is a "pre-exhumation" surface which has survived erosion.

The extent of the exhumation can be gauged from the massive buttes and cliffs which stand alone along the North Platte River, and which so amazed the pioneers along the Oregon Trail. Such are Chimney Rock and Scotts Bluff. The latter stands 200 m180m/600 feet above the river.

River systems resolve themselves into a number or easily-recognisable patterns: the dendritic, characteristic of flat regions with broadly uniform lithology; parallel, on steeper slopes where several rivers can flow out in parallel; trellis, a rectangular pattern which forms in areas of synclines and anticlines with frequent river capture; and radial, where the drainage emanates from a dome, as in the English Lake District.

After Davis, geomorphologists have had little problem with the idea of structure or process. The division into stages has however proved very controversial. One problem is that the simple situation of rapid uplift of a planed surface seems rarely to happen, as rapid erosion begins after any uplift. There is no obvious peneplain anywhere in Britain, and moreover there is scarcely a major valley in the country which does not show signs of repeated changes in level, marked by knick points and paired sets of terraces. The base sea-level has altered far too frequently, interrupting one cycle of erosion and starting a new one, for anything like a peneplain to form. Places of great geological stability – millions of years of little change, as far back as the Mesozoic – notably in Africa and Australia, do have vast, flat plains, so the concept is not without value. In dry areas a different form of erosion takes place in any case, already discussed as pedimentation and scarp retreat.

It appears that some landscapes simply show an adjustment to structure and the idea of "stage" – youth, old age – seems irrelevant. For example in the south of England are two adjacent chalk areas, Salisbury Plain the Downs, which display completely different slope angles. Under the Davisian system, steep slopes represent youth, and flat ones old age. However Salisbury Plain is formed from almost horizontal strata, and is flat, whereas the Downs consist of steeply folded anticlines and synclines, and have steep slopes. Both are of similar age and have the same climate: stage obviously has nothing to do with it.

One idea which has developed to replace stage is dynamic equilibrium, where a landscape, including its angles of slope, river bed gradient and relief, stays more or less the same as it ages, and lowers at a more or less uniform rate. If there is then a change in "inputs" – in the type of weathering, for example from temperate to periglacial – then the landforms will adjust to that change. One way to envisage this sort of thing is a "rock sandwich" with two hard beds of sandstone above and below a soft clay. Rivers will cut deep beds into the resistant top layer of sandstone, but when this is removed by denudation, the same rivers will then form broad valleys in the clay. When in turn this is removed the rivers will form steep valleys once more in the second bed of sandstone.

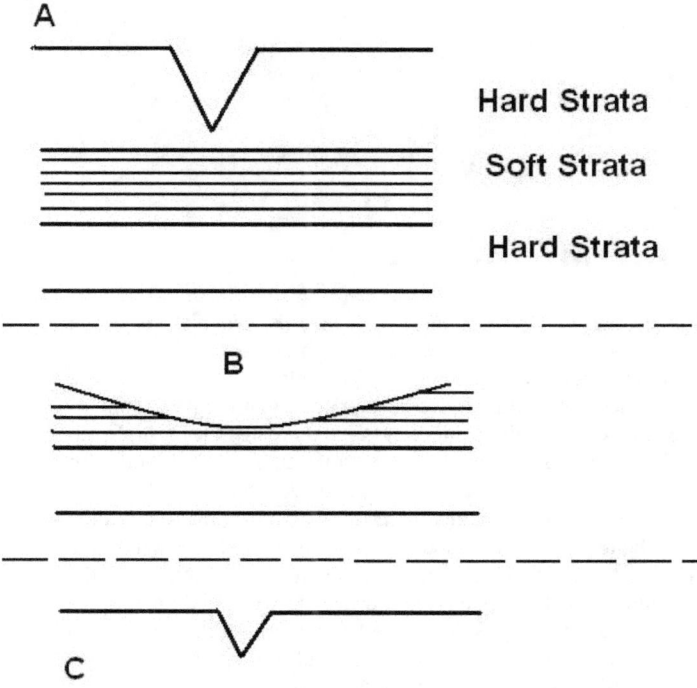

Dynamic equilibrium where the slope profile is sharp, then concave, then sharp again as the different layers of rocks are eroded away

****

It is difficult to imagine a peneplain forming in the completely unstable conditions of Britain. There is widespread evidence from abandoned strandlines that the country was inundated to a depth of 180 metres (585 feet) at the start of the Quaternary – the Calabrian transgression – yet during the depths of glaciation, the sea level was probably lower by a similar order of magnitude. However, some rather obvious levels do "shout" from the map or from the landscape, indicating that peneplains may once have existed but are now dissected. Geomorphology people spend a lot of time looking at this kind of thing. The summits of central southern and south-eastern England cluster around similar heights, 215-275 metres (700-900 feet) above sea level. A large part of Wales running between Snowdonia in the north to the Brecon Beacons in the south lies at 450-600 metres (1450-1950 feet) above sea level, sloping gently to the south, and has been recognised as a platform of erosion since the middle of the nineteenth century. There is speculation about the age of these platforms – the Welsh one may be very old, in fact Cretaceous, with a layer of chalk now removed, as in Yorkshire.

****

I spent the first part of my working life as a land use planning officer in the Eastern Province of Zambia, which has an area about the size of Wales and a population at that time (1973) of only about half a million. I hadn't been there long when it became apparent that the geography of Africa does not work in the same way as it does in Europe or North America. Half the province consisted of "plateau", uplifted land with little by way of mountains, standing at about 900 metres (3000 feet) above sea level. This contained almost all the population, which favoured the better agricultural land based on weathered fans below the hills. The other half of the province was "valley" – in this case, the Luangwa Valley. The Luangwa is a major tributary of the Zambezi. Its valley lies at about 450 metres (1500 feet) above sea level, so there is a big drop in altitude from the plateau. As a result it is much hotter, it is full of tsetse flies, bringer of sleeping sickness to mankind and trypanosomiasis to cattle, also full of all types of African game, and almost devoid of people. The vegetation changes, mixed acacia woodland and grassy wet dambos giving way to giant baobab and sausage trees. The valley bottoms contain many ox-bow lakes, infested by crocodiles. The Zambezi Valley itself is much the same.

As in the Eastern Province of Zambia, so over much of southern and central Africa, the valleys are regarded as pestilential, and are avoided by mankind. The major centres of population centre on the better agricultural land of the plateaus and on mining areas. In Zambia there is such a mining area, the Copperbelt, where there is a string of towns. The centre of civilization is linear, based on the "Line of Rail" heading northwards from South Africa and Zimbabwe to the capital, Lusaka, and on to the Copperbelt. (The geology of the plateau of the Eastern Province and the Copperbelt of central Zambia is all Precambrian.) The pattern is much the same in South Africa, where the big cities are centred on the gold mining area, the Rand, or the coast. In Zambia there are no cities on river junctions, as you would expect in Europe. The mighty Zambezi flows almost alone through its valley. Only small towns mark the junction of the Luangwa and Zambezi. It is the same on the Limpopo, which flows out to the Indian Ocean in Mozambique, and along most of the Congo and its large tributaries. Because there are few cities, there is little river traffic, but there is another reason for that.

The uplift which has created the plateaus creates a difficulty for the rivers, which clearly must find their way down to sea level. The result is some of the world's most spectacular waterfalls. The point at which a river works its way back from its lower valley to a higher level is known as the knick point, and this is normally marked by rapids or waterfalls. On the Zambezi these are the Victoria Falls. The same effect can be seen on the Congo, where there is a whole series of falls and rapids collectively known as Livingstone Falls after the calm of Stanley Pool. 320 kilometres (200 miles) inland from the coast, these falls were a barrier to navigation on the Congo for most of the nineteenth century until Henry Stanley himself supervised the construction of a road – later a railway – past them. Apart from Kisangani (Stanleyville) on the middle Congo, it is only here that major cities are found on the river – Brazzaville and Kinshasa, opposing capitals on either side of the Pool.

In the nineteenth century the holy grail of the African explorers was to find the source of the Nile. The first white man to see the middle course of the Congo, called the Lualaba, was Dr Livingstone. As it was already a mile wide and evidently in the middle of a continent, he thought he had found the Nile itself. However, others were sceptical, because of the low elevation of the Lualaba – it was not considered to be running at a high enough altitude to flow the more than 6400 kilometres (4000 miles) to the Mediterranean. This calculation turned out to be correct. Stanley himself was the first man to navigate down the Lualaba, thereby establishing that it was the Congo and not the Nile.

His reputation has not survived especially well, but he was by far the most persistent and successful of the African explorers.

The real source of the White (western) Nile is Lake Victoria, which does have the appropriate altitude. This was discovered by John Speke in 1862. The other source of the Nile is the highlands of Ethiopia, where the Blue Nile rises. The White Nile is larger of the two by volume, but dissipates much of its water in the great marshes – the Sudd – of Sudan. It is the Blue Nile which contributes the annual flood which nourished the ancient civilization of Egypt.

I first set eyes on the Zambezi River in 1974, when a civil war was under way in what was then Rhodesia (now Zimbabwe). The Zambians had closed the border at the river, the frontier with Rhodesia, but we were going to Rhodesia! We made good progress on this trip, stopping the night in Lusaka and then heading down the Central Province into Southern Province. We gazed in amazement as we passed through the vast agricultural estates of the Southern Province, mostly farmed by Europeans, and growing maize and sugar cane – there was nothing like this in our province. As the border with Rhodesia was closed, the only way to get across the Zambezi was to drive to the point where the neighbouring country to Rhodesia, Botswana, touched the river on the opposite bank.

Here there was a pontoon, an unpowered wooden barge that you could drive onto. There was a cable strung across the river, and the boatmen would place wooden clubs with notches cut in them on this cable and drag the barge across, a few feet at a time. There were manned gun and mortar emplacements on both sides of the river, which was itself infested with crocodiles and hippos, and it was a long way across.

"We can't go on that!" exclaimed Christine.

"It's the only way across. We have travelled 900 miles (1440 km) to get this far and we are not going back now!"

I gingerly drove onto the pontoon, and off we went. The thing rocked in the water. It was the end of the rainy season, and the river was swollen and full of swirling eddies. This was not a journey for the faint-hearted! We clambered out gratefully on the Botswana side, cleared immigration – a rough hut – went down a sandy road for a mile or so, then it was immigration again – this time the white men of Rhodesia. Within a few days we had arrived at the mighty Victoria Falls, *Mosi-oa-tunya*, the Smoke that Thunders, one of the natural wonders of the world, and impossible to view without getting wet! We were there in April, immediately after the end of the local rainy season.

We could see the "smoke" miles before we got there. The volume of water going over the falls was terrific – and do you know what? We were practically the only tourists there.

****

At the opposite end of the temperature scale, a very different hydrology can be witnessed in Siberia, where there are examples of the powerful effects of ice and ice damming. It is one of Russia's misfortunes that its vast Siberian province has a mountainous southern border, instead of a mountainous coastal border, in the manner of the western United States. This dictates that its great rivers must flow northwards. Siberia has four great river systems, from west to east the Ob, Yenisey, Lena and Amur. Of these, only the Amur flows from west to east, draining into the Sea of Okhotsk. The others all flow into the Arctic. Of these three, the easternmost, the Lena, has the least hospitable course within a drainage area of frozen mountains which can never make a welcoming home for humans. The Ob and the Yenisey basins however contain land with more potential, but the south to north direction of flow presents a problem. It means that when the summer snowmelt season has begun in the south of Siberia, the north is still frozen, and the water cannot escape into the sea until the much later Arctic thaw sets in. The result is the biggest bog in the world, known as the West Siberian Plain. This occupies the entire area between the Ural Mountains in the west, the Yenisey River in the east, and the Altai Mountains in the south-east. It covers an area of more than 2.65 million square kilometres (975,000 square miles), a third of all Siberia, and is the world's largest unbroken plain. More than half of it lies at less than 100 metres (330 feet) above sea level. A rise of fifty metres (162 feet) in sea level – not much, by Quaternary standards – would cause all land between the Arctic Ocean and the city of Novosibirsk to be inundated.

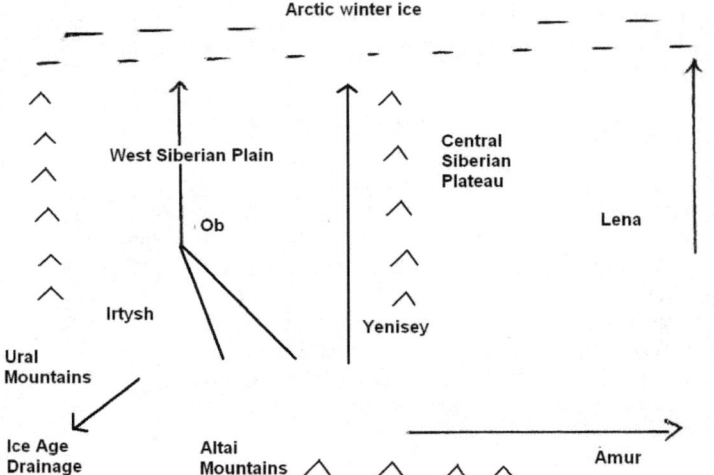

The drainage pattern of Siberia – the Ob, Yenisey and Lena all melt in the south before the spring thaw in the north

This is a region which has undergone prolonged subsidence throughout the Cenozoic. The geology is comprised largely of horizontally-bedded alluvial deposits. Many are quite recent, having formed at the bottom of periglacial lakes. During the Ice Age, the route north was entirely blocked by ice dams, and a truly monumental lakes built up, causing both the Ob and the Yenisey to reverse direction and drain south-westwards into the Caspian Sea. Russian engineers have seriously considered trying to re-engineer this feat.

However, it is not all bad news for the Russians – there are substantial oil and gas reserves under the plain, exploited from the seventies onwards.

****

The drainage of the Alps offers an insight into river capture. Anyone driving across Bavaria in the south of Germany will note the large number of big, fast rivers flowing northwards, out of the mountains and in the direction of the coast. Only one of them gets there, beyond Bavaria in the west – the Rhine, which cheats by flowing through a tectonically-constructed valley. All the other rivers, including the Isar, on which Munich stands, are captured by the Danube, which flows from west to east right across Germany on its way into the Balkans and south-east Europe. Far from starting life in the high Alps, it drains the relatively humble Black Forest, and water which was destined for the North Sea ends up in the Black Sea. This probably means that this is not a classic case of river capture. The Black Forest is a Hercynian structure, much older than the Alps. It is quite likely that a river following the line of the Danube already existed, before the uplift of the Alps, and the vigorous rivers flowing northwards from the new mountains simply flowed into it.

In other respects, the drainage of the high Alps is very similar to that of the Himalayas. The Po takes the place of the Ganges, draining the southern slopes of the mountains and flowing laterally across the foot of them, almost in parallel. The role taken by the Indus is in the Alps taken by the Rhone, which flows north-westwards away from the Alps into Lake Geneva, before turning round the flank of the mountains to drain into the Mediterranean near Marseilles, as does the Indus into the Arabian Sea at Karachi. On the eastern flank, the drainage from behind the Alps goes into the Danube, as noted. This does not break through to the Mediterranean – although it does get quite close in the Balkans. Instead, it is diverted far away to the Black Sea. In this way it resembles not the Brahmaputra, but another great river rising in the Himalayas, the Yangtze.

The parallels between Italy and the Indian subcontinent are quite striking, as they have a similar tectonic role. Not only is the drainage system similar, but so is the overall shape, with the wide submontane valleys in the north and a crumpled southern section. Each even has a substantial offshore island off the southern tip – Sicily and Sri Lanka. However, in other respects, Italy resembles the British Isles, but in place of a mountain barrier, there is a maritime one, the Channel taking the place of the Alps in dividing Britain from the rest of Europe. Again, each country has a large offshore island (Ireland). As a matter of fact the distribution of the main cities is also similar, but upside-down. London and Bristol occupy analogous positions to Venice and Genoa, Liverpool and Glasgow to Rome and Naples. Italy could be described as an island state with a barrier of mountains instead of the sea.

A similar drainage pattern to the Danube can be seen in the rivers of Yorkshire – Swale, Ure (the river of Wensleydale), Nidd, Wharfe, Aire and Calder (known by the acronym SUNWAC). These rivers form the drainage of the eastern Pennines, running from the crest of the hills at heights around 600 metres (2000 feet) to the east coast of Yorkshire. However, only one of these rivers reaches the coast, as it has gathered the drainage of all of the others. The confluence of the Swale, the Ure and the Nidd above York forms the Ouse, joined below York by the Wharfe. This river is then joined from the west by the combined Aire and Calder to form the Humber. The Calder is the only river in the system to run almost straight west to east to the coast. This is not a classic case of river capture, however. The northern rivers could never flow directly to the coast, as the high ground of the North York Moors and the Yorkshire Wolds lies in the way, and therefore they are bound by gravity to flow into the southern rivers anyway.

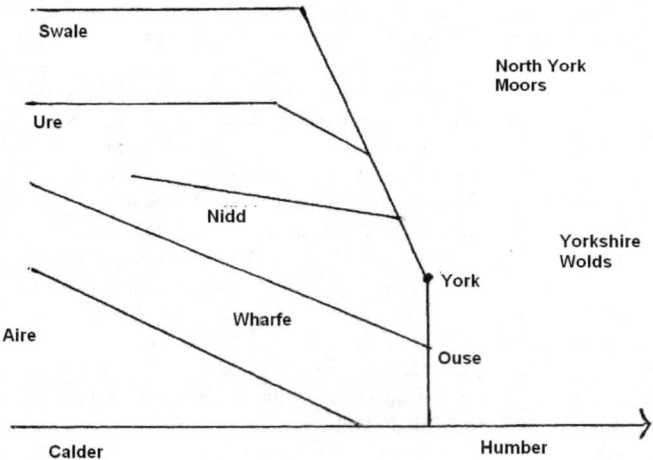

Drainage of the Yorkshire rivers – it looks like river capture by the Calder but local topography has played a part

****

Some readers might be surprised to know that the continent of Australia has anything resembling a pattern of drainage, but in fact its drainage is of a very distinctive character. Inland from all coasts except the long central section of the south coast (the Nullabor Palin where no trees grow) is an area where relatively short rivers drain to the coast. This zone is easily the widest in the north of the continent, where, however, most of the rivers have seasonal flow only. In the east of Australia is the Great Dividing Range, and to the east of that is the coastal drainage to the east coast. To the south and west of this lies the second drainage area, the Murray-Darling basin, which drains most of New South Wales and the southern part of Queensland, exiting to the coast near Adelaide in South Australia. These two drainage zones could be considered normal even in a temperate region.

The rest of the country – that is, about half of it, is occupied by two zones of desert drainage, east and west. The eastern part is slightly wetter and higher, with well-defined drainage channels terminating in salt lakes, the largest of which is Lake Eyre, which expand greatly in the wet season. The western part of the desert drainage has approached peneplain status. As explained above, Australia has drifted through a much wetter climatic zone to its current position under the diverging, sinking air of the subtropical desert zone of the southern hemisphere. It was in this earlier, pluvial period that the drainage system of the western deserts was laid down, millions of years ago, when the relief was higher and the valleys were well defined. Now all that is left is the senile remains of these valleys. The land is so flat and the climate is so dry that rainfall flows in sheets down the slopes, rather than flowing in well-defined channels, and ends up in salt lakes which dot the courses of the ancient rivers.

For all the world like a post-Impressionist painting by Cezanne, this is the **Yellowstone River** as it flows from the National Park

My final river is the Yellowstone, a fitting place to end a geology book, as the park through which it flows is one of the earth's great geological wonders. The river itself rises on the Continental Divide, which passes through a corner of the park; to the west lies the Snake River basin, where the drainage is to the Pacific; to the east, the Yellowstone flows away to become a major tributary of the Missouri, draining eventually into the Gulf of Mexico. The upper reaches of the Yellowstone, including its most striking waterfall, captured the imagination of the American public in the nineteenth century when they appeared in a famous painting. This was *The Grand Canyon of the Yellowstone*, by Thomas Moran. Having been to the park, I would say that this painting represents something of an idealised, romaticised view, but let's call it artistic licence, the view is in any event spectacular!

Of course, the river itself is not the only feature of interest. Most of the park, which sits largely in a corner of the state of Wyoming, is in fact a volcanic caldera, created by a massive rhyolitic explosion 640,000 year ago, estimated to have been a thousand times larger than the eruption of Mount St Helens in 1980. Moreover this was not the first eruption, nor is it expected to be the last, since the rumblings continue and the Old Faithful geyser blows every hour! There were other notable eruptions 1.2 million and 2.1 million years ago. The very name of the park – yellow stone – refers to the rhyolitic tuff which is found there.

The park sits at a high level, 2,400 metres (8000 feet) and is rimmed by higher peaks, the edges of the caldera. The volcanicity is due to a stationary feature of the mantle, an area of hot rocks 500 km/300 miles deep and 100 km/60 miles wide. Where this approaches the surface there is a magma chamber, which causes this surface to belly out and subside back with the passage of time. The North American plate has moved across this massive lava chamber, so that the evidence of previous eruptions now lies in the mountains to the west of the park, stretching into southern Oregon. It is known that each major eruption covered large parts of the current United States with a thick layer of ash, to the extent that any future eruption on a similar scale would virtually wipe out life over thousands of square miles.

Appendix I – Mountain Building in the USA

Although this book is mainly concerned with the geology of the British Isles, the geology of the United States offers some interesting contrasts, notably in respect of the remarkable mountain ranges which occupy the east and west of that large country, so I have included a description here as an appendix.

The Appalachian Mountains of North America, stretching northeastwards all the way from central Alabama in the south to New England, eastern Canada and Newfoundland in the north, seem to represent a rather uniform feature of the North American continent. The mountains bend first this way, then that, but the chain, as seen on a globe, does look as if it represents a single creation. Its characteristic ranges and valleys all follow a similar trend, SW-NE or WSW-ENE. Indeed there is one great valley within it which runs from Alabama to New Jersey and on into Canada, known as the Great Valley of the Appalachians, seemingly a single geological creation, and occupied along its length by various rivers, including the Tennessee, Shenandoah and Champlain.

However, geologists think that the Appalachian chain was built in three distinct and very long phases, know at the Taconic, Acadian and Alleghanian orogenies, running from 450 (or earlier) to 200 million years ago. It follows that the formations involved in the various stages of the folding were Palaeozoic – in fact very much the same sort of rocks which were involved in the Caledonian and Hercynian orogenies in the British Isles, old sandstones, limestones, conglomerates, shales and the like.

The Taconic orogeny coincided broadly with the Caledonian orogeny of Europe, and indeed the mountains of northern Scotland and Norway are regarded as broken-off pieces of the Appalachians. It is thought that, during the earlier stages of the closure of the Iapetus Ocean, the northern part of what is now the Appalachians ran into a subducting oceanic terrane carrying an offshore island arc. (A terrane may be defined as a piece of land broken off a tectonic plate and accreted onto the edge of another, normally oceanic to continental.) In the next phase, the Acadian (375-325 Mya), affecting particularly the north of the chain, it is thought that North America may have collided with a mini-continent – in fact part of Avalonia (the Avalon peninsula is part of Newfoundland). The final phase, the Alleghanian, took place 325 to 260 million years ago as Africa collided with North America during the formation of Pangaea.

It is tempting to ascribe these successive orogenies in terms of plate tectonics as first Baltica and later Africa crashed into Laurentia, but whilst this works fine on paper, it does not really match the evidence coming from American ground geologists, or from recent seismic data taken from the American mainland. For example, there is a distinct shortage of suture zones, where the converging continents conjoined. One such was thought to be the Brevard Zone, dating from the Alleghanian, and running north-east up the east coast through North Carolina. However, this is too short (about 160 km/100 miles) and has proven not to descend to any great depth, as would be expected from the seismic data. Also, if these plates were colliding, why did the whole process take so long – 250 million years? That is simply far too much time, even in geological terms.

Some geologists are now in favour of resurrecting geosynclines instead of trying to invoke plate tectonics as an explanation of these orogenies. There is evidence that only processes local to Laurentia were involved. For example a large and important formation known at the Martinsburg slates of Pennsylvania, New Jersey and adjacent states seem to have been formed largely from deposits brought from the east, quite possibly an eroding island arc such as Japan represents off Asia today. It is dated to the middle Ordovician, around 450 Mya, and clearly represents the deposit of mudstone in what used to be called a geosyncline. Subsequent compression and metamorphosis took place in the Taconic, where the general direction of movement was north-west, along thrust faults. This does not really fit with a collision with Baltica in the east. It fits much better with the idea of the docking of an oceanic terrane coming from the south-east. Some "post-plate tectonic" geologists now speculate that lightweight sediments accumulated to great depths in the Martinsburg basin, and were then subjected to mountain-building processes of unknown origin, but definitely not connected with plate tectonics.

The same process can be seen operating today, for example off the mouths of the Rhine, Mississippi or Ganges. Huge depths of sediments are forming on the sea floor, but in none of these cases does it seem likely that plate tectonics will one day raise them back up as new land. They are nowhere near plate boundaries. However some other process, as yet unknown, may do just that. Meanwhile, the creation of the Andes illustrates the process of mountain-building without a closing land mass.

The result of all this today in the Appalachians is a range of folded and faulted mountain with peaks such as Mount Katahdin reaching over 1600 m/5,200 feet – not, then, the Rockies or the Alps, which are newer ranges. To the west of the mountains lie the coal and oilfields of Pennsylvania, Pennsylvanian (Upper Carboniferous) in age. West again in Ohio are very flat strata developed on top of the central craton of North America, a place which has not witnessed an orogeny in a billion years, and which in certain eras has dipped to allow the formation of marine deposits on top of it.

By the end of the Mesozoic era, the Appalachians are thought to have been worn down almost to a plain (a "peneplain"). Uplift during the Tertiary period caused the rivers to etch out weaker beds, notably carbonates, into valleys, and leave the harder beds, again notably quartzites, as the montane ridges. It is this process which caused the famous geomorphologist W M Davis to develop his theory of landscape evolution (see Chapter 21, **Rivers**).

Geomorphologists believe that the drainage of the eastern United States was originally largely to the west, to the basins of the Midwest which accumulated layer after layer of sedimentary deposits over long periods in the Palaeozoic. The uplift of the Appalachians changed that pattern, so now the drainage is to the east coast, or to the south. For example the Delaware River now has to find its way across the Appalachian ranges in an eastward flow, as it does at the Delaware Water Gap. The Ohio River is thought to have been formed by the diversion of pre-existing drainage during very recent (Pleistocene, Ice Age) times.

****

Have you ever looked at a relief map of North America? One thing is very striking – the western mountains. The big chains pile one upon another in the west; they stretch all the way from Alaska to Mexico; and they form a very wide belt, 1600 km/a thousand miles or more, roughly a third of the width of the continent. No other mountains in the world are quite like this, but of course, there is a reason for this configuration of the continent. North America has been migrating westwards for upwards of 200 million years, since the breakup of Pangaea, overriding oceanic crust all the way, and pushing up one mountain range after another.

In a way it is simplest to start by looking at the Andes, because these mountains are more recent than the Rockies, and have a much simpler structure. To the west of them and off the coast is the Peru-Chile oceanic trench where the Nazca plate is subducting beneath South America. To the east and on land is a tectonic accretion zone, where new land has piled up against the edge of the continent. This represents sediments from the sea bed, sea mounts, abyssal hills or islands which were previously in the way. These structures are called accretionary terranes and we have met them before, when considering the Ordovician and Silurian rocks of the Southern Uplands of Scotland as mapped by Charles Lapworth. (A terrane may be more narrowly defined as a piece of land broken off one tectonic plate and accreted onto another.) East of the accreted zone is a volcanic arc, stretching the length of the continent. It lies inland because the mantle rock above a subducting plate only begins to melt when the subbducted plate is 130-160 km/80-100 miles underground. The melting is caused by the release of water from the subducted plate, which catalyses the melting process by interfering with the chemical bonds of the minerals in the mantle rock. (This surprising fact has been demonstrated in pressure ovens.) The lava is andesitic.

Eastwards again lies a fold and thrust belt where the rocks have been folded into mountains. Finally to the east once more is a fifth zone, a foreland basin, which represents an area of land pushed downwards by the immense weight of the mountains of the fold and thrust belt. This is full of sediments washed down from the mountains to the west.

The story of the Rocky Mountains in North America is similar but much older. Here the mountains began to rise about 200 million years ago, perhaps 25 million years after the break-up of Pangaea. The continent was drawn westwards, over what geologists call the Farallon Plate. Part of the accreted terrane from this includes relatively large amounts of ophiolites, which are very common in California. These consist largely of pillow lavas ploughed up from the sea bed, and mantle material (peridotite) altered by heated sea water to give it a smooth but fibrous texture and a green-black colour. The name of this rock is serpentinite, and its dominant mineral is serpentine (*ophis* is the Greek word for snake). In California it is closely associated with gold deposits. To the ocean bottom rocks were added anything that lay in the wake of the advancing plate and was not subducted, including small continents and island arcs.

Three main phases of mountain building are recognised, the main effects of which spread gradually eastwards, so that the latest phase, the Laramide, is most noted in Wyoming and Colorado – hundreds of miles inland.

The earliest and westernmost mountain building phase is known as the Nevadan, and took place throughout Jurassic times, 200-145 million years ago. In this phase, much of western California was constructed by coastal accretion, while inland lay a volcanic arc, a fold-and-thrust zone and a foreland basin, as in the modern Andes. The foreland basin, covering Colorado, Wyoming and other western states, became the site of the terrestrial Jurassic Morrison Formation, noted above under the Jurassic for the most famous dinosaur beds in the world.

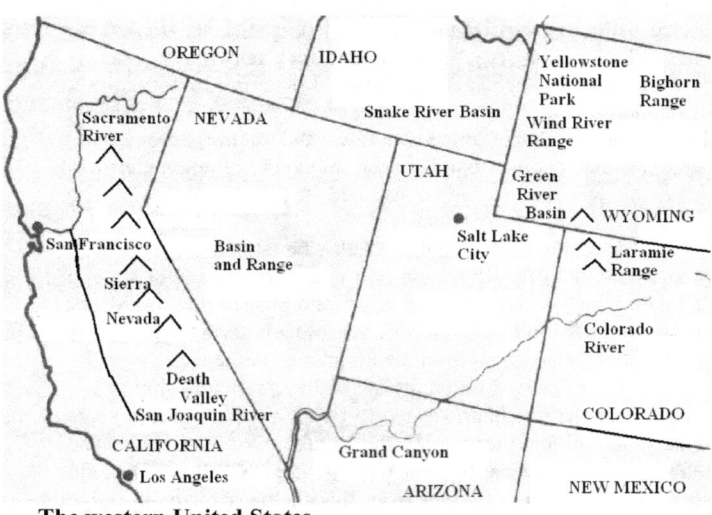

The western United States

The next phase, central in terms of both geography and time, is called the Sevier Orogeny, and dates from the Cretaceous, 145-65 million years ago. It affected what is now Nevada, Idaho, Utah, Arizona and western Wyoming. Arizona, Nevada and Idaho were the main volcanic states – in other words the volcanic arc had migrated to the east. The Sierra Nevada dates from this phase – it is a massive granite batholith, exhumed from ten or more miles down. It extends from the central Valley of California into Nevada as far as the Basin and Range Province. In the fold and thrust belt, east of this, some formations have been found which have been moved 160 km/100 miles eastwards from their point of origin along low-angle thrust faults.

The foreland basin for this orogeny lay to the east, in Wyoming, Colorado and eastern Utah. This is the home of Cretaceous dinosaurs, especially the duck-billed hardrosaurs (bipedal grazers including *Maiasaura)* and their predators including of course *Tyrannosaurus rex*. The central states of North America were at this time submerged under the Cretaceous Interior Seaway. This eventually reached a width of 1600 km/almost a thousand miles, stretched right from the Gulf of Mexico to the Arctic and lasted for 20 million years. It left behind it a thick layer of chalk and other sediments.

Beyond the mountains of the Sevier orogeny and to the east of them, the final orogeny, the Laramide took place 75 to 45 million years ago. It was this phase which gave the American West its most magnificent and distinctive mountains, known as the Foreland Ranges. These are found mainly in Colorado and Wyoming, America's highest states in terms of average elevation. They include the Bighorn, Laramie and Wind River Ranges. They have not been produced by folding and thrusting, however, but by massive vertical movements caused by the compression. The vertical displacement in the Wind River Range, where the highest peaks reach 4,300m/14,000 feet, is believed to be at least 13,500 m/44,000 feet! The same ancient basement rocks which outcrop at these summits have also been found by drilling down to 9,200 m/30,000 feet under the nearby Green River Basin. This displacement – about 13.5 km/eight miles – is thought to have taken place over a period of 30 million years. It is suggested that the root cause is that the Farallon Plate flattened out in its plunge beneath the continent, bringing horizontal pressure to bear on the rocks lying ahead of it to the east. Where this type of mountain building has taken place, there is also an absence of volcanic activity.

In fact the same phenomenon can be observed in a section of the Andes today, where subsurface flattening of the angle of subduction can be observed from seismic data. It is thought that the cause of it is a thickening of the basaltic layer of the oceanic plate (before subduction). Though heavier than most continental rocks, basalt is lighter than the typical peridotite of the top part of the mantle which also forms the bottom part of the moving plate. Therefore where the basalt layer is thick, the average density of the subducting plate is less than the mantle into which it is moving, and so it "floats" at a high level instead of diving down at an angle of 45 degrees. The reason there is no surface vulcanicity is that the plate has not dived deep enough to cause melting.

****

The landscape of much of south-western United States today owes little to ancient orogenies and a great deal to a modern one, exemplified by Death Valley in eastern California, which descends to 86 metres (282 feet) below sea level. Only 138 kilometres (86 miles) away in the Sierra Nevada is Mount Whitney (named after geologist Josiah Whitney), at 4,421 metres (14,505 feet) the highest peak in the contiguous United States. Death Valley is at the western end of the "Basin and Range" country of Nevada and adjacent western states, where mountain ranges lie alongside valleys or basins in parallel north-south rows. The Great Salt Lake of Utah lies towards the eastern end. The mountains rise to over 4000 m/13,000 feet and the basins lie around 1250-2150 m/4,000-7,000 feet above sea level, though that does not show the true extent of the faulting which has produced this scenery. One of the basins at Jackson Hole at the foot of the Grand Teton Range, right at the north-eastern tip of the region, contains a depth of 5,000 m/16,000 feet of sediment! Within the borders of this region lies a large area of inland drainage, the Great Basin. The whole structure has been determined by faulting and massive vertical movements, still taking place today. The mechanism which has created the Basin and Range structure is crustal stretching, as opposed the crustal compression which normally throws up mountain ranges.

Observed on the road to Canyonlands, Utah, this apparently crazy pattern appears to be caused by gullying through vertically bedded strata of tuff rather than by folding

Mountain ranges normally have deep roots going into the mantle, but the Basin and Range country does not; its roots are shallower than the average continental crust. It is thought to be buoyed up by a hot magma chamber, close to the surface. The amount of heat emitted from the earth's crust is about 50% above average in this region. The crustal stretching has been significant, of the order of 400 km/250 miles, roughly doubling the area of the Basin and Range territory. A block of granite which once lay near Las Vegas has been dragged 240 km/150 miles to the west to become the Sierra Nevada Mountains.

As the Fallaron Plate resumed normal thickness between 42 and 21 million years ago, volcanic activity resumed above the trajectory of its dive in the Basin and Range country. The whole area became covered in thick deposits of tuff which can be seen in roadside cuttings and elsewhere today in decorative bands of pink, ochre and gold. North America eventually came to override the East Pacific Rise – the mid-oceanic ridge which was the source of the Fallaron Plate. Cut off from its source, the old plate sank deep into the mantle. Its ghostly outline can still be seen in seismographic data as a 3,700 km/2,300 mile-long slab of rock. Its leading edge dips down almost to the core below Bermuda, and its trailing edge still lies under the American West. The hot mantle welling up behind it is thought to explain the heat and crustal stretching of the Basin and Range. Meanwhile GPS data shows that the Basin and Range country and the land to the west is being pulled away from the rest of North America along with the Pacific Plate, and there is speculation that one day an ocean will form in the middle of what is now Nevada. Los Angeles and San Francisco are actually sited on the Pacific Plate, across the San Andreas fault. The Pacific Plate and the Basin and Range country are pulling in the same direction – north-westwards – but the Pacific Plate is moving faster, hence the San Andreas Fault.

North of Cape Mendocino, the westernmost bulge of northern California, up to Vancouver, a fragment of the old Farallon Plate survives, as the East Pacific Rise has not been overridden here. This is known as the Juan de Fuca Plate, and its dive beneath the continent has produced the Cascade Range, which contains spectacular and recent volcanoes, including Mount St Helens and Mount Rainier. The latter hovers menacingly above the city of Seattle. South of Puerta Vallarta (situated at 20 degrees N in northern Mexico), a larger fragment of the old Fallaron Plate survives. Known as the Cocos Plate, it is subducting under Mexico and Central America.

The western extremity of the Basin and Range country is marked by the Sierra Nevada, considered a separate mountain range but involved in similar earth movements. It is a massive granite batholith from the Cretaceous period, now exhumed by upward faulting as the volcanoes which once stood on top of it have eroded away. It represents a very large block of crust, 640 km/400 miles long north to south and 96 km/60 miles wide, contain eleven peaks over 4,300 m/14,000 feet. The range tilts up asymmetrically so that the eastern side slopes down steeply, while the trajectory of the western slopes is much lower, sloping down gradually into the Central Valley of California, which itself represents the accumulated debris eroded from the mountains. Because of this configuration the Sierra Nevada ("Snowy Mountain Range" in Spanish) has wet western slopes, but the eastern slopes are barren and arid, lying in the rain shadow of the big peaks above and so forming part of the Great Basin, an area of inland drainage.

The Sierra Nevada may have been a mountain range in earlier times but it is certain that most of the faulting which has raised it up has occurred in the last ten million years, and major earthquakes still occur around it today, notably on its eastern flank. The effect of the faults is not only to raise the range vertically, but also to move it northwards. This is the effect of friction against the Pacific Plate, dragging the Sierra Nevada and the Basin and Range country away from the rest of the United States as described above.

The lower western slopes of the Sierra Nevada contain the Mother Lode of the gold which led to the California Gold Rush of 1849. The first gold was found in placer deposits – amongst the sand and gravel deposited by rivers flowing though gold-bearing rocks. The second source was ancient river beds, now perhaps 900m/3,000 feet above sea level on the slopes of the Sierra Nevada, also found to contain placer deposits. These ancient rivers flowed out from Nevada, which has its own gold-bearing lodes. All placer deposits, however, must have a source in the country rocks themselves, and for the recent Californian placers the main source was the Mother Lode. This runs north-south though the western foothills of the Sierra Nevada for about 320 km/200 miles, a line roughly followed by today's Highway 49 (inland from San Francisco). The gold was deposited in quartz veins. Super-hot hydrothermal streams percolated away from hot granite, dissolving silica and other minerals including gold, lead and magnesium as they did so, to cool off and crystallize at a distance. The hydrothermal streams sought out weaknesses in the country rock – fault and crush zones. The Mother Lode lies along the boundary or suture between two of the terranes which were scraped off the subducting Farallon Plate and stuck onto the western coast of what became the United States. The quartz veins which are there now date from Cretaceous times and so are nothing to do with the modern San Andreas Fault or with the recent faulting which has created the Basin and Range country.

The miners found the placer deposits, recent and ancient, far easier to work than the mother lode, as the gold could simply be sifted from the "auriferous gravels". Water hoses running at tremendous hydraulic pressures were used to blast through the ancient gravels, a process which caused so much damage to the environment that it was halted in 1884, 35 years after the start of the California gold rush.

West of the Sierra Nevada lies the Great Central Valley of California, an almost flat plain 640 km/400 miles long and 80 km/50 miles wide. This is a tectonically created trench, where the crust is thought to have been depressed by the downward thrust of the Fallaron plate in Jurassic times. Since that time it has filled up with flat-bedded sediments, Mesozoic and Tertiary. The current drainage is by the San Joaquin River, which flows northwards, and the Sacramento River, flowing south, the two joining and flowing into San Francisco Bay. These certainly bring in fresh deposits, but most of the sediments in the Great Central valley do not derive from the mountains which now stand to the east and west of it – in fact the Valley is much older than either of these ranges, which are recent geologically.

West of the Great Central Valley lie the Coastal Ranges of the western United States, which have been created by the pressure of the western edge of the continent against oceanic formations. This is a really mixed geology, an accretionary terrane of rocks scraped off the top of a descending plate, where the normal rules for the sequencing of either sedimentary or volcanic rocks are difficult to apply. The sequence runs for 800 km/500 miles along the coast, and it contains sediments from the continent such as terrestrial sandstone; marine formations including greywackes and red cherts; and various igneous rocks, especially ophiolites (including gabbros, pillow lavas and serpentinites). Metamorphosis has also interfered, to create blue garnet schists. The whole is known as the Franciscan mélange. During the Pliocene, red rhyolitic lavas and tuffs were added to this mixture in fairly short order as volcanoes erupted under the new structure, yielding fertile soils today.

Appendix II – The Geological Structure of China

Let us begin by a quick overview of the general shape of China. Although, as noted, China is about the same size as the United States, in terms of relief (height) it is a much more rugged and higher country. Only 25% of it is 500 metres (1640 feet) below sea level, compared to 60% of North America and 80% of Europe. In fact only 15.3% of the land area is considered arable (compare 19.2% in the United States).

Also, it only has a single sea coast, which gives it a very large and inaccessible interior. This coast is not very sailor-friendly. There are few good harbours, those of the south being hemmed in by mountains, and those of the north beset by deltaic swamps. Hence China's economic development tended to be inward-looking. There were certainly seafaring subcultures on the coast, but these were located beyond the mountains in the south-east, an economically peripheral area until relatively recently. (These coasts have now been integrated into the global transport system and have forged ahead of the interior.)

In terms of relief, three great subdivisions can be distinguished – the highest, the frigid Tibetan Plateau, at an average of over 4,000 metres (13,000 feet) above sea level. Beyond this lie the plateaus of the arid north and west, generally lying at altitudes between 1,000 and 2,000 metres (3,250 and 6,500 feet). These includes large desert or semiarid basins – the Tarim, the Junggar and Inner Mongolia – and the Loess Plateau of northern China. In addition are the much wetter dissected plateaus of the south-west in Yunnan and Guizhou. The third step is made up of the plains of eastern China, generally below 500 metres (1640 feet).

In terms of geology, the bedrock of China is not quite as important as it is in other parts of the world, as it is say in England or France. This is because the rocks themselves are subordinate to the folding and uplift in the west, which has created a broad structure more or less independent of the actual rocks, and to surface deposition in the densely settled north and east of the country. Here the bedrock geology is of little importance as these rocks are buried either under hundreds of feet of windblown sand (forming an unconsolidated soil known as loess), or under thousands of feet of alluvial sediments laid down by the great rivers of China – the Hwang He, the Yangtze, the Wei, the Huai (Hwai), the Manchurian rivers and so on.

Nevertheless the bedrock geology is very important when it comes to minerals, and especially to coal. The whole country is underlain by ancient (Precambrian) shield rocks, forming specifically the North Chinese and the Yangtze cratons in the north-east and south-east of the country. These rocks crop out at the surface from place to place, as they do along the border with Korea, but generally they are buried beneath younger strata. The south-east of China and much of the north are then dominated by rocks of Palaeozoic age. In the south, these have been disturbed by many later granite batholiths. In the north, and especially in the provinces of Inner Mongolia, Shaanxi, Shanxi and Manchuria, these rocks contain by far the most important coal measures in the country. Coal is found in most of the provinces of China, but in the south it tends to be sulphurous, ashy and of poor quality. (There is even some coal in the Yangtze Valley, for example at Anyuan in Jiangxi, where Mao Zedong once worked as a Communist agitator.) This gives China one of its greatest transport problems, because the supply of coal is predominantly in the north, but the demand for it – the burgeoning industries and cities of the south-east – is at the other end of the country. The south-west of the country is dominated by rocks of Mesozoic age.

Map 1 - Eastern China

The great geological event which affected China from about 23 million years ago was the construction of the Himalayas, starting in the Miocene epoch. The tectonic plate which carries the Indian subcontinent had been moving slowly northwards ever since the breakup of the great southern continent, Gondwana, during the Cretaceous period from about a hundred million years ago. Since Eocene times, fifty million years ago, the Indian plate has powered northwards, at a very fast rate for a tectonic plate – five centimetres (two inches) a year. When it finally hit an immovable object, the Asian tectonic plate, it ploughed beneath it, forcing up the great mountain arcs of southern Asia as it did so. Hence in the south-west of China there are a number of chains of very high mountains, notably of course the Himalayas along the borders of Tibet and Nepal, culminating in Mount Everest itself (8,850 m/29,035 feet), but these are by no means the end of the story. The Tibetan Plateau sits at a very high level, around 4,000 metres (13,000 feet) above sea level, with peaks rising another 12,000 feet or more above that. To the west of Chengdu (Chengtu) in Sichuan lies the Daxue Shan (Azure Mountain Wall), containing one peak – Mount Gongga (Minya Konka) – which stands at 7556 metres (24,790 feet), hundreds of miles from the Himalayas and the easternmost big peak in all Asia.

Map 2 - Western China

The tectonic plates continue to shift beneath China, and because of the density of the population, the result has been some devastating earthquakes. One of the largest of recent years occurred in 2008 in Sichuan province, 80 km/50 miles north-west of Chengdu. It is thought to have killed at least 80,000 people. Another large earthquake hit the steel city of Tangshan in eastern Hebei in 1976 and killed an estimated 250,000 people. However the worst disaster occurred in 1556 during the Ming dynasty, when approximately 830,000 people living in caves in the northern province of Shaanxi were killed. Hebei and Shaanxi are both in the north of China, rather a long way away from the Tibetan plateau. (Jia Lu estimates that 6.1 million people lost their lives as a result of earthquakes in the twentieth century alone, presumably most of them from the famines and disruption which followed them.)

Two of the great rivers of the India subcontinent originate in Tibet on the back slopes of the Himalayas. The Indus flows westwards and the Brahmaputra (in Tibetan, the Tsang-po, and shown on some maps as the Yaluzangbu) flows eastwards. Both rivers break through the mountain barrier to find their way into the Arabian Sea and Bay of Bengal respectively. To achieve this feat the Brahmaputra has cut a canyon which is 505 km/314 miles long and 6009 m/19,714 feet deep! It is likely that these rivers show antecedent drainage – that they existed before the great uplift began, and have managed to keep pace with it, as has for example the Colorado River in the United States.

However another famous river was so badly affected by tectonic uplift that it had to turn round. This is the Yangtze, in the Himalayan foothills of south-western China. The river makes every show of draining southwards, as the nearby and parallel great rivers the Mekong and the Salween continue to do, heading for Hanoi and the Gulf of Tonkin. Instead its way is blocked by a massif called Cloud Mountain at the town of Shigu in Yunnan province, and it turns back northwards, past Jade Dragon Snow Mountain, cutting the Tiger Leaping Gorge as it does so (such wonderful names!) After one more brief attempt to cut its way south, the great river drains eastwards, on into Sichuan (Szechuan). The way out of this province is blocked once again by mountains, but the Yangtze cuts through these, exiting at the Three Gorges dam, finally to reach the east coast of China at Shanghai. It was the mountain building of the Himalayan orogeny which blocked the course of the Yangtze, creating numerous earthquakes as it did so. The former course of the Yangtze southwards is still marked by a valley, now occupied by the Red River, but the biggest tributary has headed away. The writer Simon Winchester, who himself studied geology at Oxford, tells this story in his book, *The River at the Centre of the World*.

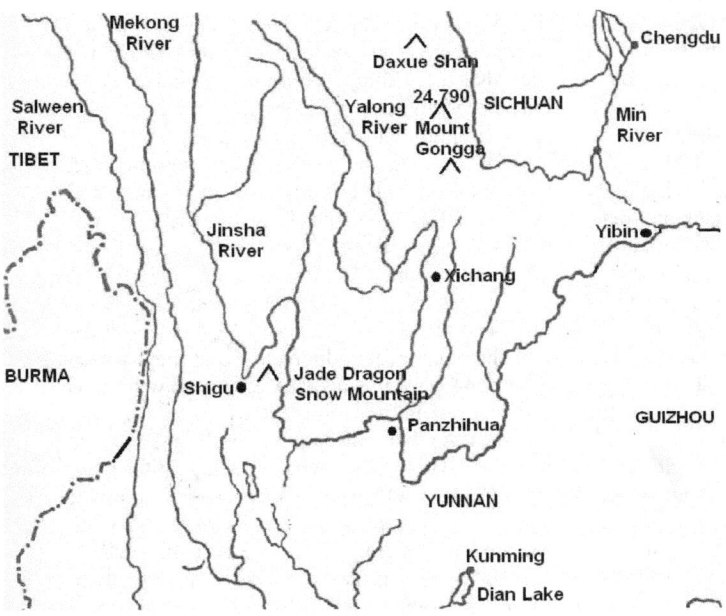

Map 3 - The Upper Yangtze, called the Jinsha above Yibin, and the Chang below it.

At 6418 km/3988 miles, the Yangtze is the longest river in Asia, and the third-longest in the world. The Nile and the Amazon are both longer, but their social, economic and cultural importance is far less. The Yangtze can claim to be the greatest river in the world. It carries 20 times as much water as the Hwang He. Its upstream stretch, above its confluence with the Min river at Yibin in Sichuan, is known as the Jinsha or "Golden Sand" River. Below that point it has another name, the Chang Jiang – "Great River". Ocean-going ships can travel up the river as far as Wuhan, 1000 km/620 miles from the mouth of the river.

Northwards into central Asia lie a series of mountain ranges which enclose desert or semi-desert basins featuring inland drainage. The largest basin, which stands out on any atlas of the world or of Asia, is the Tarim, which contains one of the world's most fearsome deserts, the Takla Makan (which in local language mans "he who goes in does not come out"!) The basin was once part of the Mesozoic Tethys Sea which largely disappeared as Africa and India collided with Europe and Asia. All that remains of it are the Mediterranean, Black, Caspian and Aral Seas. This means that the sedimentary rocks underlying the Tarim Basin were laid down in Mesozoic and Tertiary times, and they have proved to be rich in oil and gas.

To the south of the Tarim lies the Kunlun Shan, to the west, outside China, the Pamirs, and to the north the Tian (Tien) Shan. The Takla Makan serves to restrict travel to its northern and southern fringes, both of which are dotted with oases based on the rivers which arise in the mountains, only to peter out in the desert. These oases were part of the famous Silk Road, not in fact one road but a thread of intertwining routes which connected China to the west from ancient times.

To the south-east of the Tarim is the Qaidam (Tsaidam) Basin, lying between the Altun Shan to the south and the Qilian Shan to the north. North of the Tarim is another basin, the Turpan (Turfan), which descends to 154 m/505 feet below sea level, the third-lowest place on land on the earth. There is a further large desert basin to the north again, the Junggar (Dzungarian), lying between the Tian and the Altay (Altai) mountain ranges. The westernmost part of Dzungaria, beyond Ürümqi (Urumchi), is known as the Dzungarian Gate, a valley which provides the only low-level route through the chain of mountains which extends all the way from Afghanistan to Manchuria. Lying well to the north of the regular Silk Road, and today on the border between China and Kazakhstan, it was more often used by migrating nomads than by merchants as a passage between the steppes of central Asia and inner Mongolia.

Lying to the east of The Tarim and Junggar Basins is yet another area of inland drainage surrounded by mountains, Inner Mongolia, which comprises the southern half of the Gobi Desert. This descends to about 460 metres (1500 feet) at its centre. A number of rivers descend from the Qilian (southern) Shan across the "panhandle" of Gansu (Kansu) to lose themselves in the trackless wastes of the Gobi. Gansu is the province which lies between Xinjiang and Inner Mongolia to the east; it narrows to its "panhandle" in its central section, where the modern highway and railway skirt the foot of the Qilian mountains south of the Gobi.

All the western mountain ranges, plateaus and basins in Tibet, Xinjiang (Sinkiang) and Inner Mongolia are very sparsely populated, though of great interest to explorers. One of the best travel books ever written about China, Peter Fleming's *News from Tartary*, deals with his journey across the Qaidam and the Tarim basins. However within

China Proper there are also mountain ranges which, if at lower elevations, have had a much greater impact on the development of China. Of these, the principal is the Qinling (Tsingling), running to the south of the ancient capital, Xi'an, which separates the drainage of the Hwang and Wei rivers to the north from the Yangtze and its tributaries to the south. This is a fundamental dividing line in Chinese geography, separating the colder, drier north from the hotter, wetter south. The very people themselves are physically different either side of this range. To the north the people are browner and taller.

There is one more mountainous area of China, the south-western plateau. This is a highly dissected area which occupies all of the province of Yunnan and the western part of Guizhou (Kweichow). Adjacent to Burma, it is the one great plateau which actually drains into the sea. Its humid valleys, thousands of feet deep, have historically proved very difficult to cross. It was across this plateau that the Burma Road was built before the Second World War.

After the mountains and plateaus come the great river basins of China. The first of these is the plain of the Hwang He. This lies between the Yin Shan of Inner Mongolia, to the north, and the Qinling Mountains to the south, then stretches off eastwards across the plain of north China until it reaches the coast in the Gulf of Bohai. Historically, the river has been very wayward, and has changed its course many times, most notably by draining to the south of the Shandong peninsula instead of (as now) to the north of it. The distance between the two mouths is a matter of 480 km (300 miles). The river itself is historically associated with disaster in Chinese minds.

Emerging from the Qilian Mountains, the river turns north in a great loop around the Ordos desert. It then turns abruptly east, then back southwards before making another lurch left-handed below its confluence with the Wei River. The middle part of its course cuts through the loess region of windblown sands which give the river an exceptionally heavy load of silt.

Between the basins of the Hwang and Yangtze Rivers lies that of the Huai (Hwai) River, covering an area one and a half times the size of England in Henan, Anhui and Jiangsu provinces. 150 million people live here. This river flows into a series of lakes, the largest of which are the Hongze and Gaoyou, north of Nanjing, then finds its way by multiple distributaries across the plains to the sea.

Of greater agricultural importance than the plains of the Hwang He today are the plains and basin of the Yangtze River, south of the Qinling axis. A third of the course of the river lies in its mountainous descent from Tibet, but it enters the plains in the densely populated inland Red Basin of Sichuan, where it picks up four large left-bank tributaries, including the Jialing River at Chongqing. It then passes through the low ranges of the Hubei Mountains, containing its three sets of gorges, before issuing out into a second plain, the lake-studded Central Basin of Hunan, Hupeh and Jianxi (Kiangsi) (note that Jaingxi means "west river"). Here it picks up a number of important tributaries including the Han River on its left bank. Finally it heads eastwards for the coast in the delta region in the third of its great plains.

The Yangtze Gorges

The third river basin is that of the Xi, in the south of the country. The Xi River rises in the mountain jungles of Yunnan and flows west through Guangxi and Guangdong before it enters the sea in an extensive delta at Guangzhou (Canton), where it is known as the Pearl River. Though not on the scale of the Yangtze, the Xi River is still 2200 kilometres/1375 miles long and flows through a very wet area. The plain is relatively narrow and supports double cropping with its subtropical climate. The most famous valley in this area is that of the Li River, where the local limestone has weathered into fantastic conical hills, a great draw for the artists of old China and tourists today. This scenery is best displayed between Guilin and Yangshuo in Guangxi. The Li ultimately flows into the Xi.

The final plains area lies in Manchuria in the north-east of China. Here are broad alluvial plains surrounded by high mountains. To the north, the Songhua (Sungari) River drains into the Heilongjiang (Amur) on the Russian border. To the south are the plains of the Liao River. The northern valleys of the Songhua and the Heilong are low-lying, cold and very boggy, but even here, in the Sanjian Plain, the drainage of the marshes is taking place. China is just so short of arable land – there is barely a hectare of it per person, which is very low by world standards.

Finally in the south-east of China, between the valleys of the Yangtze and the lower reaches of the Xi and its eastern tributaries, is an area of dissected, folded mountains. This has a noted regional trend, south-west to north-east. The rivers are short and fast-flowing, disgorging into the East and South China Seas and the Taiwan Strait. This area has a long coastline, lying between Ningbo and Shantou, and its people long resisted incorporation into China Proper. It regional dialects are incomprehensible to other Chinese. The area traditionally looked to the sea for a livelihood, and emigration, especially from Fujian, has long been seen as the only solution, rather as in Ireland. However it is this area which has recently attracted much of the new industrial growth. The area includes several old treaty ports, including Amoy, now known as Xiamen.

****

I have some personal experience of the Yangtze River. The enthusiast can always find some geology on his holidays! I went with my wife on a boat cruise down the river, from Chunking, the largest city in Szechuan, two and a half days down the river to the Three Gorges Dam. The waters from the dam back up to Chunking and beyond – in fact a total distance of 600 km (375 miles). The construction of this dam has caused the rapids though the mountains behind it to be drowned and effectively made safe for navigation. But how much has been lost here? Until recently the river narrowed to 45 metres (fifty yards) wide, and there is a tradition of boats being hauled up through the worst of the rapids by straining, naked men. Those days have now gone, and with them much of the attraction of the gorges themselves. The 45-metre stretch is now over 180 metres (200 yards) wide. At its deepest, the dam banks up 175 metres (570 feet) of water, so the walls of the gorges are also much shallower and less spectacular than they used to be.

Many of the locals are certainly as sceptical as the rest of the world about the new dam. In fact it is principally a hydro-electric project. They worry about its unknown ecological effects, such as landslides occurring in future years as the local geology is undermined by rising groundwater. Then there is the danger from earthquakes or fault movements, by no means unknown in the region. The dam itself cleans the river of silt so it will emerge downstream with greater incisive power.

Some ecological effects are already known. First, the river is full of rubbish, both vegetable and man-made (polystyrene and flip-flops, mostly). The local authorities along the route try to clean it up, but the task is evidently beyond them. In the old days most of this rubbish was smashed to smithereens in the rapids or disappeared somehow, and never bothered anyone. Secondly, Chungking, already a furnace in the summer, has become even hotter. House prices in the much more pleasant upstream city of Chengdu are much higher than in Chungking, and that premium is increasing.

Along the route of the rising river, there is very little agriculture, as it passes through mountains all the way. However there are many towns and these have all had to move uphill. The locals do not mind this much as the old blocks of flats have been replaced by new ones with better facilities, though the older people fret about the graves of their ancestors. Altogether, 1.3 million people have had to be rehoused.

We found the dam at last, crawling with tourists and – worse – exceptionally importunate tradeswomen. Once at the dam our ship entered the top of the locks. There are five of these, of ship-canal size, 280 metres (910 feet) long and many metres deep. With the ships crammed in like sardines, we slowly descended over a period of just over three hours. The engineering is all very impressive, massive concrete walls and steel lock gates. But the dam itself will be at risk if one day the local geology should take its revenge for the presumption of mankind.

The Yangze Gorges

Appendix III – Geological History of Australia

The formation of Australia echoes in many respects the formation of North America, as the continent, at first small, was built up from Archaean cratons which were constructed before the start of the Proterozoic 2.5 billion years ago. The rocks in the cratons are characteristically granite and greenstone (which is metamorphosed basalt, tuff etc). The eastern half of Australia did not exist in the Archaean.

The largest of the three Archaean cratons is the Yilgarn, which occupies most of the south-western corner of Australia behind Perth. This is 1000 km/625 miles from north to south, and 600 km/375 miles from east to west, so it represents a substantial area of shield rocks. It includes the famous gold (and also nickel) fields of Kalgoorlie and Norseman. The region is characterized by granite ridges. The granites were once thought to be batholiths extending to great depths, but modern seismographic data shows that they are only about 4-7 km/2.5-5 miles deep, and that they appear to have been injected into the existing country rock as elongated pods.

The second craton is the northern Pilbara, in the north-western corner of Western Australia behind Port Hedland. This is about a third of the size of the Yilgarn. Smaller still is the Gawler craton of South Australia, extending northwards from the tip of the Eyre Peninsula inland from Spencer Gulf, the "nick" in the coast of South Australia.

Amongst the greenstones of these shield rocks is an early type of basalt called komatiite, which is unusually rich in magnesium and formed only in the Archaean when the crust was hot. This weathers out to a ridged pattern of long, thin pyroxene and olivine crystals called spinifex texture after the tough, sharp grasses of the Australian outback.

The three Archaean cratons probably represented separate islands in the ancient oceans, and may in fact have been widely separated at those times. During the next phase, the Proterozoic, the two largest combined to form a large part of what is now Western Australia. There are more, extensive Proterozoic outcrops stretching from the very centre of Australia around Alice Springs, north-westwards to the coast at Kimberley and northwards to Arnhem Land. Others are found around Mount Isa in Queensland. The Gawler craton in the south was extended eastwards to Broken Hill in New South Wales. Hydrothermal fluids permeated some of these rocks, leaving behind large bodies of mineral ores, mainly in the form of sulphides. Both Mount Isa and Broken Hill have been very important mining areas for copper, lead and zinc, with deposits dating from this era.

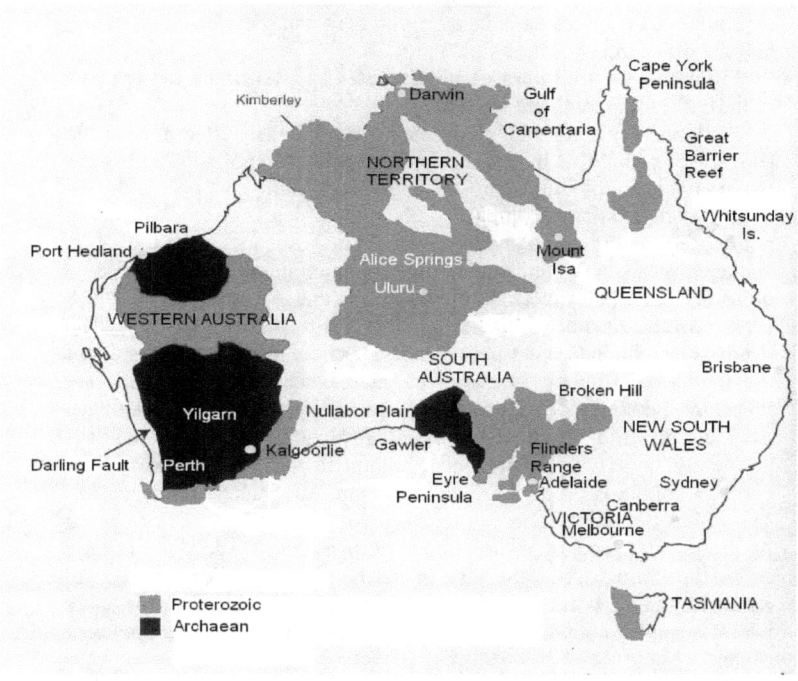

Ancient Rocks of Australia

There were signs of life in the Proterozoic, mainly in the form of stromatolites. These became very widely spread in Australia and elsewhere in the middle of the Proterozoic, before grazers evolved to feed on them. They are still famously found today at Shark Bay in Western Australia, a bay of the ocean which receives limited oceanic water and so has become too salty for stromatolite grazers.

The period 2.6-1.9 billion years ago witnessed the large-scale deposition of banded iron formations in Australia, as in Canada and elsewhere. The probable mechanism by which these were deposited is discussed above in the chapter on Precambrian rocks. The main location for them in Australia is the Hamersley Range in the Pilbara region which occupies the northern part of Western Australia. These ores can contain over 60% iron oxide and as such form one of Australia's main exports today, the main market being the steel factories of China – China does have its own iron ores, but they have only about half the iron content of the Pilbara ores.

From about 1.3 billion to 1 billion years ago, Australia formed part of the early supercontinent Rodinia.

Snowball earth, for which evidence was previously cited in Namibia and Spitzbergen, also makes an appearance in Australia. There were two main glacial phases, at around 780-700 million years ago, and 610-575 million years ago (compare the Namibian dates, given above as in phases between 750 mya and 590 mya). The sequence is just the same, and equally baffling – undisputed tillites, sometimes very thick indeed – up to 5.5 km/18,000 feet! The evidence is that Australia lay within 8 degrees of the equator at this time, so this really is Snowball Earth. The deposits also include wind-blown sands, such as form the loess soils of the modern ice ages. Once again intercalated with the tillites are limestones, dolomites and also ironstones, such as might very well have been laid down in a tropical sea – Paul Hoffman's carbonate caps. It is absolutely mystifying, indicating a complete and repeated switch of climatic conditions, but the evidence from Australia is undisputable. The tillites occur in scattered patches across a broad swathe of central Australia, stretching from the Flinders Range east of Adelaide in South Australia, through Alice Springs in the Northern Territory, and on to Kimberley in the northern part of Western Australia. Outside this band, which occupies about a quarter of the continent, there are no ancient tillites.

Two other things of note in Australia date from the late Precambrian, Ayer's Rock and the Ediacaran fauna. Ayer's Rock or Uluru is one of the world's most impressive monoliths. It is composed of arkose, a feldspar-rich sandstone probably originally laid down as an alluvial fan. It is thought to have been exhumed from beneath later deposits and now stands out amongst flat sand dunes and other sandy deposits of recent origin. It may seem unusual to those accustomed only to temperate landforms, but monoliths are common in tropical climates.

The Ediacaran fauna originally discovered by Reg Sprigg in 1946 is discussed above in the chapter on the Precambrian. It was discovered in the Wilpena Pound in South Australia, a broad fold of sedimentary rocks laid down 650-545 million years ago, that is, immediately before the start of the Cambrian era at 542 mya.

The first mention of the supercontinent of Gondwana in the stratigraphic history above is in the Ordovician, but Australian geologists now believe that a form of it has already assembled by 520 mya. The supercontinent was to have a very long life, finally breaking apart only after 180 mya. The early Australian contribution to this excluded eastern Australia, which was built up by later volcanic activity and sedimentation. A line known as the Tasman line separates Precambrian from the rest of Australia, passing southward-westwards from the foot of the Cape York Peninsula in Queensland in a wavy line, bulging east to take in Broken Hill, and ending at Kangaroo Island off the coast of South Australia near Adelaide.

One piece of evidence for the early existence of Gondwana is a fossil sponge-like creature called *Archeocyatha*, shaped like an inverted cone, open end upwards. This is found in large numbers in early Cambrian limestone, indicating a marine transgression. A lump of limestone containing these fossils was picked up by a member of Shackleton's first Antarctic expedition, one Frank Wright, in 1908, and the same fossil subsequently turned up in the northern part of the Flinders Range of South Australia.

During the lower Palaeozoic, there were two distinct geological situations in Australia. The western and northern regions of the continent were generally above water, enjoying a balmy climate in which terrestrial deposits such as river gravels were laid down. This area formed the northern coast of Gondwana. There was obviously some tectonic activity, including the first movements along the Darling fault behind Perth, which is still observable today as a scarp east of the city (the western boundary of the Yilgarn craton). From time to time, parts of the crust would subside, creating embayments of the ocean where fossiliferous limestones were deposited.

To the east of the Tasman line were shallow and then much deeper seas, containing volcanic island arcs and subject to compression and the folding of sediments laid down in the basins, much as happened in the Caledonian and Taconic orogenies of Europe and North America. Repeated orogenies are identified. These seas contained a fauna typical of the Lower Palaeozoic, and indeed there is a celebrated site for the collection of Cambrian trilobites at Beetle Creek, near Mount Isa in Queensland.

By the time of the Ordovician (488-444 mya), an interior seaway known as the Larapinta Seaway had opened up between the northern and southern parts of the west country (the east being largely under deep water at this time). This resembles the Cretaceous Western Interior Seaway of North America.

The mix of rocks laid down in the eastern basins corresponds to contemporary sequences in Britain – limestones, shales, sandstones, sometimes mixed in cyclothems, sometimes subsequently folded in the several orogenies which have been identified. There are also turbidites. Also similar is the presence of deep-water Ordovician black shales, as identified by Lapworth in the Southern Uplands. These include graptolite fossils which can be used to divide the rocks into zones deposited over less than a million years. These shales are black because the environment in which they were deposited lacked oxygen, which meant that organic matter within them was preserved to colour the rocks. They also include pyrite grains which in some cases have mineralized the graptolites.

The Ordovician sequences of central and northern Australia are relatively undisturbed, but in south-eastern Australia, extending from New South Wales southwards to Victoria and Tasmania, they were compressed with a north-south trend. The Ordovician in this area was terminated around 440 mya by the Benambran Orogeny, an east-west compression which created major folds in central Victoria and New South Wales. In these areas, Silurian sediments lie above crumpled Ordovician sedimentary rocks unconformably.

The pattern continued into the Silurian and Devonian. By the time of the middle Silurian, the eastern part of Australia was substantially formed, and the island arcs lay 300 km/180 miles further east, in fact due north of the position of Sydney. An ocean lay off the north-west coast of the continent, containing depositional basins. The red sandstones exposed along the gorge of the Murchison River in Western Australia date from this era. They contain footprints of giant water scorpions. The earliest plant remains found in Australia date from the early Devonian (412 mya; the Devonian began 416 mya). Ferns and primitive gymnosperms which could grow into trees were common by the late Devonian. The first land animals also appeared – centipedes, spiders and amphibians.

Associated with the repeated orogenies of the Lower Palaeozoic in the east of Australia is the emplacement of granites. These cover considerable areas of the south-east of the country notably around Canberra. One such is the Cooma granite, 114 km/70 miles south of Canberra, and which is surrounded by severely distorted gneiss. Their dates range from 435 mya but emplacements continued over many millions of years, up to 90 mya.

The last of the Lower Palaeozoic orogenies was the Devonian Kanimblan (387-370 mya) which marked the changeover from the balmy times of the Lower Palaeozoic into the icy conditions of the Carboniferous and Permian. The great ice age of this era is thought to have fundamentally the same cause as the Pleistocene ice age – the drifting of continental areas of land over the south pole, where no warm sea current could melt the winter ice.

During the Carboniferous, Australia resembled Alaska today – semi-polar, covered in ice sheets and glaciers, and also volcanic with an offshore island arc, in the manner of the Aleutian Islands. The continent formed the north-eastern and coastal part of Gondwana, with Antarctica conjoined and to the south, very close to the south pole. There appears to have been some uplift and erosion during the Carboniferous, so that not much remains in Australia in terms of Late Carboniferous sediments. During the Permian, as the ice thawed, deposits were left behind on top of a glaciated landscape.

The Carboniferous period began by 359 mya but ice conditions did not set in in earnest until about 330 mya. When the ice ages came, they seem to have been very similar to the recent ice ages, in that there was a vast but fluctuating spread of ice sheets. At the same time there was active volcanism in an island arc which stretched northwards for 1800 km/1125 miles from Sydney into Queensland. There were large outpourings of rhyolitic lavas, and tuff.

There are three main types of evidence of the Permo-Carboniferous glaciation of Australia today. The first is the widespread presence of striated and polished pavements, especially in the southern parts of South Australia, Victoria and in Tasmania, that is, the parts which were nearest to the ice sheets spreading out of Antarctica. The second line of evidence is the presence of tillite and erratic boulders. Thirdly there are dropstones in marine and lake deposits. Some of the dropstones and erratics are very large and obvious.

Within South Australia the best examples of glaciated landforms are in the Fleurieu Peninsula, south of Adelaide.

The ice sheets came and went over millions of years, reaching a late maximum at 275 mya, and still persisting after that. Palaeosols (fossils soils) from Kiama on the coast of New South Wales dated to 258 mya show permafrost-style features as seen today in the frozen soils of Alaska and Canada.

In the seas, the warm-water fauna of the previous era, including corals, calcareous algae, molluscs and brachiopods, was replaced by a cold-water fauna with a different set of brachiopods and molluscs, plus bryozoans (sea mats) and echinoderms (sea urchins), and definitely no coral reefs.

The Late Carboniferous and more especially the Permian was a great time of coal accumulation in Australia. The source of the coal was not the tropical swamps which created the coal of Europe, but cold-water peat bogs. These filled very extensive areas, as they do today in western Siberia, the greatest modern bog in the world. In these cold climates, the peat accumulates very slowly, at a rate of approximately 1mm a year, and is at first 90% water. It is thought that a coal seam 1m thick would have taken 10,000 years to accumulate, and to produce good clean coal, it must in that time not have been contaminated by sediment. Even so, there are coal seams 9m/30 feet thick in Australia. Most of the coal basins line up down the extreme eastern side of the country, all the way from the base of the Cape York Peninsula to Melbourne. The best-known coal-mining region of Australia is the Hunter Valley, north of Sydney, and centred on the city of Newcastle. The beds here contain a celebrated collection of fossilised Permian insects, caught in a shower of tuff, at Belmont. However there are many coalfields and today a string of ports to service them along Australia's eastern coast, including Abbot Point and Hay Point in Queensland, which export a large proportion of the world's seaborne coal.

The dominant vegetation of the Carboniferous peat bogs was however the same type of primitive spore-bearing tree ferns as in the tropical areas. These were lycopods, including species of *Lepidodendron*, still with its characteristic Pirelli tyre-track bark pattern. During the Permian this was replaced by the southern-hemisphere *Glossopteris* flora. These plants are early gymnosperms, that is, seed-bearing ferns.

During the earlier part of the Permian (299-251 mya) there was also a marine transgression, consistent with the melting of the ice sheets. Many coasts were inundated and there were major marine embayments in four places: to the east of Adelaide and into Victoria and New South Wales; in the eastern part of the Nullabor Plain in the south; in the north-west, north of the Pilbara, where the Great Sandy Desert now lies; and in Tasmania. By the later Permian only the last two of these were still under water. The volcanic activity which had continued for so long in a belt north of Sydney had ceased, to be replaced by another marine inundation in the Brisbane region. The end of the Permian was marked by the great mass extinction found around the world at that time.

By the start of the Triassic period, 250 mya, two things had happened. The enormous, cold Permian peat bogs had dried out as the climate grew much drier – to the point of aridity – and warmer. Also Australia had moved directly over the south pole, which was located in western New South Wales at 240 mya. However, the ice had all gone, and the landscape was dominated by inland drainage and lakes. The aridity is thought at least partly to be due to the formation of Pangaea, of which Australia formed the southern part, by the middle Permian. Sediments were principally of two types, alluvial sandstones and red mudstones – in fact red rocks dominate the Triassic in many parts of the world, including England (New Red Sandstone) and similar facies in the United States. The red-brown mudstones, known as "redbeds", have been much quarried to make the red brick houses of Sydney. By the middle Triassic, 230 mya, the climate had become wetter and peat bogs had started to form once more.

The vegetation of the wetter areas was typical of the era: conifers, ginkgos, lycopods, ferns and horsetails,. A large herb-like plant called *Dicrodium* formed widespread heaths. The fossil record of the mainly terrestrial rocks is understandably poor, as it is in the New Red Sandstone formed in similar circumstances in Britain. However it is clear that the Triassic in Australia looked much as it did elsewhere. An almost complete large salamander-type amphibian called a labirinthodont, 2.25 m/over 7 feet from head to tail, has been recovered from redbeds near Sydney.

The eastern coastal regions of Australia from northern Queensland to south of Sydney in New South Wales formed a depositional basin in Permian and Triassic times. This extended inland to two further basins, the Galilee in Queensland, and the Cooper to the south of it, which extends into South Australia. The southern part of the coastal region is the Sydney basin, which is 350 km/220 miles across, and which is filled with sediments from these times. At first (early Permian) these were predominantly marine, then later mainly terrestrial, and included layers of coal. There are also some intercalated volcanic rocks, probably from volcanoes lying off the east coast. Further evidence of volcanism is found in Tasmania, where there is a part of an enormous dolerite sill known as the Ferrar Dolerite. This is also found in South Africa and Antarctica and is thought to have a volume of 2.5 million cubic km/600,000 cubic miles. Dated at 182 mya, its presence suggests that the crust of the earth was starting to move in Pangaea, and that magma was rising from the mantle to fill in the gaps.

In the Jurassic (199-146 mya), the climate of Australia grew generally wetter. Still there were peat bogs which eventually created coal, mined behind Brisbane in Queensland in the Rosewood-Walloon area. The main contributor to the coal was a tree similar to *Araucaria* (the kauri pine). There were even dinosaurs at these southerly latitudes. The largest-known Australian dinosaur was a creature called *Rhoetosaurus*, which could reach 17m/55 feet in length and weigh 20 tons.

If anyone is wondering how these typical Mesozoic plants and animals, including amphibians, dinosaurs and monkey puzzle trees, ever reached Australia, well, it was part of Pangaea, and therefore part of a continuous land area which stretched north of the Equator and included most of the continents. The mystery is how a dinosaur – a reptile dependent on the Sun for heat – could support itself in conditions where there were two months in the year with no sunshine! Seasonal migration is possible, though a journey of thousands of miles would not seem practical for a 20-ton dinosaur. However there has been a long debate amongst palaeontologists as to whether the dinosaurs WERE reptiles, as they bequeathed us the birds, which are warm-blooded. Yet a warm-blooded creature without any form of insulation would find it difficult to survive an Antarctic winter even in the Jurassic.

During the later stages of the Jurassic, India split off from Gondwana to begin its long journey to Asia, creating a new western coast for Australia along the lines of the present coast. The part of India which used to be here has either been lost entirely, or has disappeared under the Himalayas. (For millions of years, the plate carrying India was a separate entity to the one carrying Australia and New Guinea, but the two now appear locked together and are known as one plate, the Indian-Australian.) Also around this time there was underground igneous activity in western Australia. This included the injection of pipes, some of them containing diamonds, which are mined at Argyle in the Kimberley area today – in fact Argyle is the largest producer of diamonds by volume in the world, and the only producer of pink diamonds.

During the early part of the Cretaceous, the same thing happened in Australia as in North America and Europe – the greatest marine inundation of the Phanerozoic. By 117 mya, half the continent was under water. The parts which were left above water included the Yilgarn craton of western Australia, the Kimberley area of the north-west, extending into central Australia, the volcanic area of the east coast, and most of the south and south-east coast. The ocean formed a funnel, with the narrow end at the foot of the Cape York Peninsula and the broad end stretching roughly from the Nullabor Plain to the east coast at Sydney. This area included most of modern Queensland and large parts of the Northern Territory, South Australia and New South Wales.

The surface subsided slowly over what must have been very flat land under the weight of the incoming ocean. The sediments which formed in this shallow sea were washed down from the remaining uplands, so this was not chalk territory. Firstly they were sandstones, then clays – a fact of the greatest later significance, because the sandstones are permeable, and the clays are not. Much of this area now forms the Great Artesian Basin, where underground water is tapped to provide most of the water available to farmers and homesteads over this vast area – 1.7m square km/660,000 square miles. Without this artesian water, most of this area would be unusable to agriculture, as the area to the west of it still is. The water which falls on northern Queensland and on the western slopes of the Great Dividing Range, which runs right down the eastern coast of Australia, finds its way into this gigantic aquifer, and moving very slowly, in some cases over thousands of years, percolates under the very centre of the continent. The sea had gone by 99 mya.

Notably between 125-120 mya there was intense volcanic activity in a new chain of volcanoes, the East Australian Arc, which stretched all the way from northern Queensland to Tasmania. Most of the vast outpourings of andesitic lava and tuff from these volcanoes has since eroded away, but some of it remains, most notably to form the famous Whitsunday Islands which rise above the Great Barrier Reef, and on the southern coast of Victoria. This is yet further evidence of the nature of the eastern provinces of Australia, subject to repeated cycles of sedimentation, orogenies and volcanism right from the start of the Cambrian, the coast extending gradually eastwards all the time. The andesitic and rhyolitic volcanic material indicates a subducting oceanic plate. This must have represented sea floor spreading from some distant mid-ocean ridge in the ancient oceans. This is not the same case as western North America, where the continent itself overrode the oceanic plates, and gradually consumed more and more of them by the accretion of oceanic terranes. It is also notable that though there were many orogenies, as in the eastern United States, there was not always a closing land mass on the opposite shore – the same situation as has now been postulated for the Appalachians. (During the assemblage of Pangaea, there would have been a closing landmass.)

Australia had a warm, wet and temperate climate during the earlier part of the Cretaceous, supporting a flora with predominant araucarian pines, maidenhair trees and podocarps (pine-like trees without cones). Modern versions of these trees are still found in the rain forests from New Guinea to Tasmania. The undergrowth contained mosses and ferns. By 110 mya the first angiosperms (flowering plants) appeared in Australia, and have been identified from the coal measures at Maryborough in Queensland. Dinosaur fossils from this era are relatively common, and include the bones of giant grazing sauropods and also the predator *Allosaurus*.

An extraordinary lake deposit from Koonwarra in Gippsland, Victoria dating from 118-115 mya contains the well-preserved remains of many creatures including fish, lungfish, molluscs and insects. There is one unmistakable feather in the deposits, yet no other bird fossil is known in Australia until 30 mya, a gap of 85 million years, which just shows what a haphazard business fossil preservation is! (Note however that *Archaeopteryx* is 150 million years old.)

The Cretaceous seas and estuaries contained a similar fauna to that found in Europe, including ichthyosaurs, pliosaurs, plesiosaurs and ammonites. One noted plesiosaur is a monster predator called *Kronosaurus queenslandicus*, and this could be 17m/55 feet long. There is a museum devoted to the display of Cretaceous marine reptiles at Richmond in central Queensland, close to the area where many of their fossils have been found.

At around 99 mya, a narrow tectonic split is thought to have appeared between Australia and Antarctica, resembling the Red Sea now. This slowly widened and has now reached 3000 km/1800 miles. The Australian plate has been the great mover, ploughing inexorably northwards into Indonesia, and creating volcanic havoc there and in New Guinea; Antarctica has stayed more or less where it was in the Cretaceous, moving slowly to its position plum over the south pole by the late Oligocene. Splitting off with Australia was an adjacent and very large piece of crust now known as the Lord Howe Rise, which used to lie along the south-east coast of Australia, and which now sits in the middle of the ocean, half-way between Australia and New Zealand. The only land here is the small island, Lord Howe Island, which seems something of a loss to the world somehow, as a South Sea island the size of Greenland would come in handy. (Note, however, that the island is a volcano and stands 2000m/6500 feet above the sea floor, so this piece of continental crust will not be resurfacing at the next marine regression!)

Much of Australia has remained above sea level for millions of years, and indeed there are surfaces which are thought to date back 150 million years! Though eroded since then, their old form remains. During Tertiary times, what is now the inland drainage basin of the eastern deserts – which now drain into seasonal lakes, notably Lake Eyre – received vast amounts of terrestrial sediments both from the west and the east, but most notably from the east. 50 million years ago, this was not a region of salt lakes, but a huge area of freshwater lakes and billabongs, much as Kakadu (a nature reserve in the north of Northern Territory) is today. There are fossils of freshwater dolphins, crocodiles, flamingos, possums and proto-kangaroos here.

The same type of deposition occurred in the Murray-Darling basin which lies to the east and south of the Lake Eyre drainage. Boreholes have found sediments dating back 50 million years in the Murray basin. Another type of deposition then occurred in the Murray Basin, which formed an embayment of the sea in later Tertiary times, somewhat resembling the London Basin. The line between terrestrial and marine deposits advanced southwards, then back northwards. However there are Tertiary limestones, marls and sandstones here which form the basis of some of the best agricultural land in Australia – for once, it is land which has not long been leached of its nutrients – laterite iron pans 150 million years old have been found elsewhere in the country!

Another embayment of the sea occupied the area of the Nullabor Plain along the south coast, giving it a coating of Tertiary limestones dated before 15 million years ago (Oligocene and Miocene). These are exposed at the surface today, and have weathered out to fantastic caves with stalagmites and stalagtites in earlier pluvial periods. Today the Nullabor is a grassy plain (its name means "no trees" in Latin) and a calcareous dust blows off it into South Australia.

****

Basaltic volcanism is a feature of the Great Dividing Range, with dates going back about 70 million years and extending almost to the present – the last eruption was only 4,600 years ago in South Australia, obviously away from the main range. This volcanism is of quite a different type to that which helped build eastern Australia, which was andesitic or rhyolitic, and appears to have a connection with the construction of the Great Dividing Range itself. There are two types of eruption, one of lava fields with spreading from multiple fissures and vents, where the age of the fields is haphazard from north to south. The other type features a central peak with lava spreading outwards, downhill, to form a Hawaiian-style shield volcano. The ages of this second type show a distinct sequence, from 33 million years old at Cape Hillsborough in Queensland to 6 million years north of Melbourne. This age range fits remarkably well with the speed of the drift of the Australian plate northwards, so a stationary mantle plume, in the manner of the Hawaiian plume, has been suggested. However there are problems as the whole picture is complicated by the lava fields which do not fit this pattern. In addition, there are two more, parallel chains of volcanic sea mounts off the east coast, known as the Tasman and Lord Howe sea mounts. These also age from south to north, with the younger ones lying to the south.

The term "Great Dividing Range" is something of a misnomer because, although the chain of mountains reaches a height of 2228m/7241 feet at Mount Kosciusko in the south, it rather fizzles out in the north, where it is only a few hundred metres above sea level. This does rather suggest that it is in some way buoyed up by the volcanic activity associated with the basaltic central volcanoes, in that as this activity has ceased in the north, the mountains have sunk back down. Again, there are large areas of negative magnetic anomalies under the chain, much more extensive in the south than the north, which suggest the deep roots of old mountains.

The basalt appears rather more interesting than the basalt of Iceland because it contains significant quantities of precious and semi-precious stones, including sapphires, rubies, diamonds, zircons and peridots. It is far easier to win these stones from later alluvial deposits than from the country basalt, as the erosion and deposition has made them much more accessible and concentrated in placers. The main areas for commercial and amateur extraction of these stones are at Anakie in Queensland, and in the New England (north-eastern) and Crookwell-Oberon regions of New South Wales. The outpouring of lava, especially in the south, also form the basis for some very productive soils.

Two of the sea stacks known as the **Twelve Apostles**, Victoria, Australia, cut in Miocene limestone

During the Oligocene and the next epoch, the Miocene, Australia drifted through a much wetter climatic zone than its current position on the globe can offer, and the result was a rich, Amazonian-style flora and fauna. Some of this has been preserved in marvellous detail at Riversleigh in northern Queensland, one of those few sites in the world where every detail of the animals has been preserved. This revealed many types of previously undreamt-of marsupials. There was even a meat-eating kangaroo. Exhibits from Riversleigh, dating back 25 million years, are on display in a museum at Mount Isa, also in Queensland.

This era has left behind it brown coals, found especially in Victoria but also in Western Australia, formed from peat bogs.

However, all the while the continent was drying out as it passed northwards into the zone of air subsidence and divergence. Deserts began to develop. As recently as 12 million years ago there was a sharp falling-off in the pollen of *Nothofagus*, a type of beech which is today found is the rain forests of New Guinea and Tasmania. At the same time there was a sharp increase in the amount of eucalyptus pollen and the recorded incidence of natural fires (ash zones). One writer (David Johnson) claims that it cannot be a coincidence that Australia dried out as its old neighbour, Antarctica, headed south and accumulated vast ice sheets just at the same time. Well I would like to state my view that this IS a coincidence. There was still plenty of water left in the oceans to make rain for Australia! The fact is that Australia had moved into the zone of subtropical air subsidence, the place where the Sahara Desert and Arabia are also found. Only in unusual circumstances do these zones become pluvial, and they go all the way back to the Old Red Sandstone of the Devonian, at least! The cooling air of the Southern Ocean would have been a factor in reducing humidity and so precipitation, however.

The oldest of Australia's modern animals is the duckbilled platypus – an opalised jaw of this type of animal has been found and dated to 110 mya! Rather famously, Australia later developed its very own megafauna. The marsupials were the dominant mammals, including the huge *Diprotodon* or giant wombat, thought to weigh over two tonnes. This animal was preyed upon by *Thylacoleo*, a marsupial lion! Then there was *Megalania*, a gigantic goanna which could reach 6 m/20 feet in length, enough to make the Komodo dragons (which make life such fun on certain islands of Indonesia) look small.

Suddenly, at 45,000 years ago, there was a mass extinction of the modern examples of this megafauna, just as there was in North America at 11,000 years ago, and in South America a thousand years later. This can only mean one thing – all of these extinctions coincided with the spread of mankind across these continents. 23 of the 24 genera of large animals (weighing over 45 kg) in Australia became extinct.

The ice ages of the Pleistocene came and went in Australia as elsewhere, with glaciers forming even on the mountains of New Guinea and Tasmania. The general dryness of the glacial periods contributed to the increasing aridity of the continent. At glacial maxima the coastline extended to the very edge of the continental shelf, then withdrew as the ice sheets melted.

****

It is well-known that the seas off Australia are very dangerous places for swimming, due to the presence of many poisonous jellyfish, vicious sharks and marine crocodiles. However, did you know that the Australian outback is also the worst place in the world for geologizing? This is due almost entirely the ubiquitous *Spinifex*, composed of nothing but a mass of silica-tipped spines. It would be more fun falling on a bed of nails than bumping into one of these things, but it happens all the time! Of course you will probably have to drive to the field site first, and that is not without its attendant dangers either. In all my years of driving, I have never been given a breath test in Britain, or indeed anywhere in Europe. However, it has happened to me on average once every week or two in Australia. It seems to me that after the morning rush, the local police have time on their hands, so they set up road blocks to persecute innocent motorists. However they have never caught me yet!

Appendix IV – Zambia Notes

Most of this country, landlocked and lying in the middle of the southern part of Africa, is built up of Proterozoic rocks. As the Proterozoic covers two billion years, that clearly means that there are many formations involved. Dating from the later part of the Proterozoic is the Katangan Supergroup, stretching across the northern and north-western parts of the country. It is these rocks which contain the copper and cobalt deposits which have dominated the economy of Zambia for so long (and also the copper deposits of the Katanga district of the Congo).

How do the copper ores come to be concentrated in this way? In fact they are located in marine deposits, but the ores themselves are associated with volcanoes. They are mainly found in the form of porphyry, a type of igneous rock which consists of large-grained crystals such as feldspar or quartz set in a matrix of much finer feldspar or quartz. These form in the roof of a magma chamber which lies beneath an active volcano. As the magma cools to form granite, metal-rich aqueous solutions are expelled from the chamber. These fluids percolate upwards through the volcanic vent, which has already solidified into porphyries. Only specific temperatures and pressures allow the ores to concentrate into ores which contain concentrations large enough to be worth mining. In fact it takes about 50,000 years for a magma chamber to expel all of its fluid, and during this time, the precipitation zone barely moves. This enables substantial copper deposits to form about two kilometres/just over a mile beneath the surface of the earth.

Outside the Zambian capital, Lusaka, most of the main towns are in the mining area – the Copperbelt. Younger rocks termed the Karoo sedimentary succession, dating from the Carboniferous to the lower Cretaceous, dominate the lower territory of the Zambezi and Luangwa valleys. The large western province of Zambia is different again, having a surface cover of late Tertiary to Pleistocene Kalahari sands. It is across this territory that the upper Zambezi River flows.

When I went to Zambia with my new wife in 1973, it was to Chipata, capital of the Eastern Province, which bordered Malawi to the east. We stayed there two and a half years, and for the first year at least, we didn't have much fun at all, and frequently considered returning home, but we stuck it out in the end. I worked as the Provincial Land Use Planning Officer in the Department of Agriculture – in fact it was quite an important job, with its own spending allocation. There was certainly a geological element to it, but this concerned agriculture, not mining. My wife also worked, as a schoolteacher.

It was a very long way to any half-decent shopping in Zambia – in fact 600 km/375 miles down a good tar road to Lusaka – but Lilongwe in Malawi had better shopping than Lusaka, and that was only 48 km/30 miles away, albeit across an international border and a very broad but poor dirt road. So it was that we often went to Malawi, where we could go further to the excellent beach resorts of Lake Malawi, notably Salima. However in the dry season – June to October – our main recreation area was the Luangwa Valley, which lies on the western boundary of the Eastern Province.

The Eastern Province is divided into two regions, plateau and Valley – the Luangwa Valley. The plateau has reasonable rains (about 75 cm/30" a year) and is relatively high – about 900m/3000 feet, which means that the climate is quite pleasant, and never that hot or sticky. The Valley is lower, drier and much hotter. This is also the land of the game reserve. The Luangwa Valley is one of the least well-known and frequented game reserves in Africa, but that is only because it is so inaccessible. In terms of what you could see, it was one of the best, and probably still is. The boundary of it lay only 145 km/90 miles from Chipata, down a rough and corrugated dirt road. This led through the farmlands of the plateau and across a line of hills in a stony pass, then it was down into the Valley. Quite suddenly, the vegetation changed dramatically. There were gigantic, ancient baobabs, scrubby acacias, and "sausage" trees, so called because their giant pods are shaped like frankfurters, though each one must weigh about eight pounds. Also there were always wild guinea fowl to be seen along the route.

The Valley country is also the home of the tsetse fly, a carrier of sleeping sickness – "tryps" or trypanosomiasis – in humans and cattle. The human form of "tryps" is not like malaria, being comparatively rare – you have to get bitten a lot to catch it. The first time I had a bite from a tsetse fly, on the abdomen, it came up like half a melon, but it was never so bad again. An attempt was made to maintain a zone cleared of all trees a mile wide to restrict tsetse fly movement – they fly by hopping from tree to tree. When driving back from the game reserve, our car would invariably be stopped by the armed guards of the Tsetse Control, who would spray inside it with insecticide. Cattle cannot be kept where tsetse flies are found, so two thirds of the Eastern Province was unavailable for any kind of cattle farming because of them.

Each dry season, at the end of May or early in June, the houseboys from the expatriate community from Chipata would build shelters on one bank of the Luangwa River, across from the game park proper. The banks were quite high at this point – maybe 3m/ten or twelve feet, and the muddy river flowed below, perhaps 30m/a hundred feet across. The river formed the boundary of the game reserve. The "shelters" were in fact quite open – they had low walls with no roofs, made of dried grasses and bamboo, and they afforded no protection at all against wild animals.

The first time we went to the expatriate camp, we were amazed at the noise at night – hippos grunting mainly, but all sorts of other screeches and roars as well. It was scary and we asked our acquaintance Mike Fox about it: "Is it safe here? It doesn't sound it."

"It's perfectly all right."

The park itself was full of wild animals – hippos, elephants, lions, zebra, giraffes, buffalo, all sorts of antelopes, warthogs, wildebeest, hyenas, wild dog, crocodiles, thousands of birds such as fisheagles and carmine bee eaters. One soon becomes an expert in the different types of antelope found here – tiny ones like duiker; then the common puku; the large, impressive grey kudu; the even larger eland, weighing up to a ton; and roan and sable antelope, also large. There were rhinos, but these were difficult to see, and also leopards, but I never did see one of those.

The roads were narrow, with the earth pushed away to either side, such that passing another vehicle was difficult, and turning round virtually impossible. I remember once driving on one such road when suddenly we were confronted with a huge bull elephant, with enormous tusks – and he was stood on the road. I stopped the car, reversed it back a foot or two. We waited, beads of sweat forming.

"What do we do now?" asked Christine.

"If he wants to charge, there was nothing we can do."

"How fast can this car reverse?"

"Not fast enough!"

Well he eyed us up and down, stood his ground, and then slowly, so slowly ambled away. Phew!

There were many prides of lions in the Luangwa Valley. First choice for dinner for a lion was buffalo. This may seem surprising, given that a buffalo is a very dangerous animal, and it has a massive plate of bone across its forehead with a sharp pair of horns attached. It is also a very heavy beast with a big kick. I think the reason that lions went for these beasts rather than something like zebras or pukus is that they are a lot slower and less agile, they gather in large herds with plenty of targets, and one animal provides a big meal. However, I wouldn't care to be the first lion to jump on top of one. The young lion hunters are known to be nervous of them, and many are tossed in the air or gored by this fearsome foe.

One of the most famous game people in Africa was established in the Luangwa Valley, a man by the name of Norman Carr. While we were there, he set up a safari camp on the banks of the river, and Christine and I went up there in an official capacity. The huts were brought in, prefabricated. I didn't like the look of them, because they were made of wood.

"Won't the termites get them, Norman?"

Anything made of wood and left outside in Zambia will be eaten by termites in very short order.

"I think we shall be all right," he replied, doubtfully.

Our camp was directly on the Luangwa River. One day, someone's dog went down to the river bank to drink. There was a sudden disturbance in the water, out came a crocodile and in went the dog, never to be seen again! This did not surprise me, because the crocodiles did not recognise the boundary of the game reserve, and they were everywhere, lying on sandbanks on either side of the river. Crocodiles did snatch children from the village dams and for humans they were more dangerous than lions. But they were not the only problem. One day the houseboys, who used to come along to cook, caught some fish in the river and left them out to dry, right outside the "shelters". In the morning, they were gone – and hyena footprints were seen in the sand. The hyena could just as well have breakfasted off human body parts taken from live humans.

An old African game ranger, quite a well-known man and an assistant to Norman Carr, was snatched by a lion while sleeping in his tent. Another was killed by a buffalo, always considered the worst of the big four – elephant, buffalo, hippo and lion – because one on his own would often charge without warning. When in a herd they were not so dangerous. The hippos could be problematical out of the water. A South African tourist on Lake Malawi was strolling along the beach at the Grand Hotel in Salima, a place we often visited, when he got between a hippo and the Lake. It gave him a bite as it went past. He did not survive.

So after being quite blasé about the game, we became extremely wary. In our last year, when visiting Luangwa Valley, we were so scared that we slept in the car, and never went again.

Appendix V – Outline Geomorphology of the Soviet Union

I have selected the former Soviet Union as the geographical unit for this appendix because, running up against the southern mountains of central Asia as it did, it makes a more coherent unit than modern Russia. Although in popular perception, Russia itself is a land of vast level plains, there is still room for great variety in an area as large as the old Soviet Union. This can notably be observed in the rugged peaks of its southern and eastern borders, higher than anything in the entire western hemisphere, running from regions on the Black Sea reminiscent of the French Riviera to the volcanic landscapes of the Kamchatka peninsula in the Far East.

In the same way the Great Britain can be divided geologically by the Tees-Exe line, separating the lowlands from the highlands, so can the Soviet Union be roughly cast into two parts, by following the River Lena, easternmost of the three great rivers of Siberia, from its mouth in the eastern Arctic to its source near Lake Baikal, then westwards 480 kilometers/300 miles from the southern border all the way to the Black Sea. North and west of this line, the relief is normally low - under 1850 meters/6000 feet - and this area covers two-thirds of the Soviet Union.

Ancient shield or craton rocks underlie the great Russian plain, but they only crop out at the surface in limited areas, most notably in the north-west of Russia in Karelia (a continuation of the Baltic shield) and the Kola peninsula to the north of it.. This type of crystalline rock also forms the sill over which the River Dnieper falls in the southern Ukraine, the site of a large HEP project. The granite and mica schists are rich in iron, manganese and nickel. This is also the case with the rocks of the Ural mountains, and which yield as great a variety of metals as anywhere else on earth. Forming the suture which linked proto-Europe to proto-Asia, they started to rise in Carboniferous times, 250 million years ago, and are still surprisingly high for such an old chain.

The newer, sedimentary rocks of the great Russian plains - the usual mixture of limestones, sandstones, clays, shales, chalk and so on - are rich in oil and gas, notably in the north of Western Siberia, which produces 70% of Russian oil today, a fact which underpins the entire modern Russian economy. Historically the most important oil region has been the Baku/Caspian area of Azerbaijan, spilling over northwards into Russia at Astrakhan.

There are also very extensive Carboniferous coal deposits, notably in the Donets basin of the Ukraine (Donbas) and further east, into central Siberia in the Kuznetsk basin (Kuzbas), and to the north of there around Karaganda.

To the south and north-east of the old Soviet Union lie the Alpine folded mountains, forming a great store for hydro-electric power. Because these mountains are so recent, they are still subject to violent earthquakes and volcanic activity. The town of Ashgabat, capital of Turkmenistan, was destroyed by an earthquake in 1948. There is a whole series of mighty mountains along the southern borders, including the Caucasus, Pamir, Tien Shan and Altai which lie at least partly within the former Soviet Union. The Pamir, also known as the "roof of the world", cover much of modern Tajikistan.

Volcanic eruptions are confined to Kamchatka, a world of its own as an icy, fiery peninsula jutting into the Pacific, and which contains one volcanic peak at over 4900 m/16,000 feet, and also the Kurile Islands, offshore in the same area and stretching down towards Japan.

The legacy of the ice ages is widespread in the old Soviet Union, though it is recognized that one of the coldest parts of it, northern Siberia east of the Tamyr peninsula (reaching out into the Arctic and blocking shipping) was not glaciated because it was too dry. The great Siberian rivers, the Ob and the Yenisei, were blocked in their northward flow, which created a great lake which spilled over into the Caspian Sea. In fact the Black, Caspian and Aral Seas became one, a temporary recreation of the Tethys ocean. There have been many plans to recreate a waterway along this route, providing the Caspian with an outlet to the ocean!

Within the Russian plain today, north of a line from Lvov (near the Polish border) to Perm near the southern Urals is the glaciated and mostly forested county, and to the south is the unglaciated and generally unforested region. Some notable terminal moraines have been used as routeways, and indeed carried both the soldiers of Napoleon and Hitler to Moscow via Smolensk. In the Ukraine there are vast areas of windblown loess, another legacy of the ice ages.

Nearly half the area of the old Soviet Union is affected by permafrost, which can extend thousands of feet into the earth.

The river network of European Russia was fundamental to the very origins of the modern country, forming a radial pattern with its hub near Moscow. Low portages and later canals connect these rivers, assisting

Moscow to its preeminent place amongst the cities of Russia. The rivers of European Russia are noted for their low left banks and high right banks, a phenomenon caused by the spin of the earth itself, whereby all moving bodies in the northern hemisphere tend to be deflected to their right (Ferrel's Law). Rivers affected by this include the Dnieper and Volga. Both of these have been converted into strings of lakes by dams constructed since 1939.

However, in general, the old Soviet Union has some of the longest, highest-capacity and useless rivers in the world. This is especially true of Siberia, where the Ob/Irtysh, Yenisei and Lena all drain from south to north, so being subject to annual flooding on an enormous scale as the ice melts in the south long before it does so in the north. The other great Siberian river, the Amur, flows west-east, along the Chinese border but even this then turns the wrong way - northwards - below Khabarovsk as it nears the Pacific Ocean. Of these rivers, the Yenisei is much the largest in terms of volume, and much the most useful. The Lena drainage lies in north-eastern Siberia and is simply frozen for much of the year. The Ob meanders endlessly through the greatest swamp on earth in western Siberia, taking 1900 km/1200 miles to fall 90 m/300 feet! The Yenisei, on the other hand, drains from Lake Baikal, and is a great source of HEP. Both the Ob and the Yenisei drain into the Kara Sea, the "right" (western) side of the Tamyr Peninsula, which sticks out so far into the Arctic that it blocked all attempts at through passages for centuries. The Yenisei marks a major divide within northern Siberia, as to the west lies non-permafrozen but glaciated land, and to the east highlands of hard rock, deeply dissected, permafrozen but mainly unglaciated. The timber found along the Yenisei is a much higher grade than that of the Ob and forms one of the main cargoes down the river.

The main south-north Siberian rivers derive their maximum flow in spring when the ice melts. The Amur, on the other hand, follows the monsoon climate blowing over from China, and has its maximum flow in July and August. It is also ice-free for five months a year, but is nevertheless too shallow for much useful navigation for much of the year, the winter lows being 6-9m/20-30 feet below the summer maxima. The river also has a sand bar only 3.7 m/12 feet below the surface at it mouth.

The central parts of the old Soviet Union feature inland drainage. The most important river here is the Volga, flowing into the Caspian Sea, and much used for navigation purposes, as it has been for long periods in Russian history. The Caspian is the world's largest inland sea, with an area half as large again as the combined Great Lakes of North America. Much of it is only a third as saline as sea water, partly due to the shallow evaporating lagoon of Kara-Borgaz-Gol, almost as salty as the Dead Sea. The construction of upstream reservoirs has caused the level of the Caspian to fall, leading to a shrinkage in the area of the best fishing grounds.

The most notorious inland sea is the Aral, which has shrunk to a fraction of its former size. Its feeder rivers, the Syr Darya and the Amu Darya, have been so extensively used for irrigation purposes that there is little water left now in the Aral. This was known to be a likely effect even before the major push to irrigated cotton fields, but the projects went ahead anyway.

As with the rivers, so with the coastline of this subcontinent - it is one of the least useful in the world. In the whole area of the north, from Murmansk in the west to Vladivostok in the east, there is only one ice-free port, Murmansk itself (famously used during the Second World War because of this). The Russian border with China near Vladivostok in the Pacific coincides almost uncannily with the edge of the winter frozen ocean. The Black Sea offers much better prospects, with the Crimea, Riga and Odessa affording major ports, a feature shared by the enclave of Kaliningrad in the Baltic.

BIBLIOGRAPHY

Charles Darwin, The Voyage of the Beagle, Penguin Classics (2003) (first published 1839)

The Hidden Landscape, Richard Fortey, Pimlico (Random House) (1993)

Life, An Unauthorized Biography, Richard Fortey, Flamingo (1998)

Trilobites, Richard Fortey, Flamingo (Harper Collins) (2001)

The Earth, An Intimate History, Richard Fortey, Harper Perennial (2005)

Minerals and Rocks in Colour, J F Kirkaldy, Blandford Press (1968)

Mapping the Deep, Robert Kunzig, Sort of Books (2000)

Cassell's Atlas of Evolution, D Dixon et al, Cassell (2001)

Democratic Ideals and Reality, Halford Mackinder, Faber Finds (2009) (first published 1919)

Adam Sedgwick, Colin Speakman, Broadoak Press (1982)

The Map that Changed the World, Simon Winchester, Penguin (2001)

The River at the Centre of the World, Simon Winchester, Penguin (1996)

Walking with Dinosaurs, Tim Haines, BBC Worldwide (1999)

Walking with Beasts, Tim Haines, BBC Worldwide (2001)

Snowball Earth, Gabriele Walker, Bloomsbury (2003)

Gaia – A New Look at Life on Earth, James Lovelock, Oxford University Press (1995)

The Revenge of Gaia, James Lovelock, Allen Lane (2006)

Architects of Eternity, Richard Corfield, Headline Books (2001)

Your Body, the Fish that Evolved, Dr Keith Harrison, Metro (2007)

The Aquatic Ape Hypothesis, Elaine Morgan, Souvenir Press (1997)

The Study of Landforms, R J Small, Cambridge University Press (1978)

The Weather Makers, Tim Flannery, Penguin (2005)

H2O A Biography of Water, Philip Ball, Phoenix (1999)

Textbook of Physical Geology, G B Mahapatra, CBS (1994)

Homo Britannicus, Chris Stringer, Penguin (2006)

The Beautiful Basics of Science, Natalie Angier, Faber and Faber (2007)

Introducing Geomorphology, Adrian Harvey, Dunedin (2012)

Your Inner Fish, Neil Shubin, Penguin (2009)

Ice Age, John and Mary Gribbin, Allen Lane (2001)

Homo Britannicus, Chris Stringer, Penguin (2006)

The Malay Archipelago, Alfred Russell Wallace, Beaufoy Books (2010, originally 1869)

The Geology of Britain, Peter Toghill, Airlife (2012)

Granite and Grit, Ronald Turnbull, Francis Lincoln Ltd (2009)

Hard Road West: History and Geology along the Gold Rush Trail, Keith Meldahl, University of Chicago (2007)

Annals of the Former World, John McPhee, Farrat, Straus and Giroud (1998)

The Geology of Australia, David Johnson, Cambridge University Press (2009)

The River at the Centre of the World, Simon Winchester, Penguin (1996)

INDEX

Acadian Orogeny 290
Acanthostega 124-6
Accretionary Wedge 104
Agassiz, Lake 239
Agassiz, Louis 12, 13
Age of the Apes 214
Age of the Earth 49
Age of the Fishes 123
Aglaophyton 127
Ailsa Craig 209
Aire Gap 142
Alexander II 157
Alleghanian Orogeny 152, 291
Allosaurus 175, 183, 324
Alpine Drainage 284, 285
Alps 214
Alston Block 138
Alum Bay 206
Alvarez, Luis 194
Alvarez, Walter 194
Alvin 49
Ambulocetus 205
Ammonite 123, 147, 157, 163
Amphibian 134
Amur 337
Anamalocaris 99
Anchiornis 172
Andes Structure 293
Andesite 27, 28
Andrew, Roy Chapman 204
Andrewsarchus 204

Angiosperm 190
Anglian Glaciation 235
Anning Mary 10, 180
Antarctic Ice 213
Anteosaurus 158
Anthracite 148
Anticline 108, 198
Apatosaurus 174, 183
Appalachians 110, 290
Aquatic Ape 233
Aral Sea 338
Araucaria araucana 134
Archaea 81
Archaean 84, 85
Archaean in Australia 314
Archaeopteryx 16, 172
Archeocyatha 317
Archosaur 164
Arid Landforms 266
Arkose 31
Arran 191
Arthur's Seat 139
Ashgabat 336
Ashmoleon Museum 179
Asthenosphere 20
Augite 20
Australia 188
Australia/Antarctic split 203
Australopithecus afarensis 232
Avalonia 104, 109
Avebury Stone Circle 198
Ayer's Rock 31, 316
Bagshot Heath 206

Bagshot Sands 206
Baltica 109
Banded Iron Formation 87
Banded Iron Formations, Australia 315
Barton Formation 206
Baryonyx 193
Basalt 24
Basilosaurus 205
Basin and Range 296l 297
Bath 177
Bats 204
Beagle HMS 15
Belemnite 173
Bembridge Marls 206
Ben Nevis 129
Benambran Orogeny 318
Bering Strait 225
Beringia 225
Birds 198
Black Cuillins 209
Black Forest 153
Black Sea 338
Black Smoker 49
Blue John 21
Blue Lias 31, 176, 185
Blue-green algae 80
Bone Wars 183
Bornhardt 248
Borrowdale Volcanics 117
Boulder Clay 242
Bowen, Norman 170
Boxgrove Man 236
Brachiopod 135

Brachiopod 98
Brachiosaurus 174
Brahmaputra 220
Breckland 223
Brevard Zone 291
Broken Hill 314
Brontosaurus 174
Brontothere 204
Brotherton formation 96
Bryce Canyon 199-201
Bryophyte 127
Buckland, William 13, 180
Bunter Sandstone 166
Burgess Shale 99
Butte 268
Cadeby formation 96, 166
Cairngorms 113
Calabrian Transgression 223, 280
Calder Valley 185
Caledonian Orogeny 109, 110
California, Central Valley 299
California, Gulf of 271
Camarosaurus 174
Cannock Chase 166
Cap carbonates 89
Cape Cod 256
Carbon dioxide, Ice Age 229, 231
Carbonation 245
Carboniferous limestone 24, 119, 136
Carlsberg Ridge 46
Carpathians 216
Caspian Sea 336, 337
Cassiterite 24

Castle Rock 139
Castleton Caves 258
Catastrophism 11
Caucasus 217
Chaine des Puys 152
Chalcedony 20
Chalk 187, 206
Challenger Deep 38
Challenger HMS 38
Charnia 93
Charnwood Forest 93
Charpentier, Jean de 12
Cheddar Gorge Caves 258
Chemical erosion 245
Chenjiang 100
Chesil Beach 179, 263
Cheviot Hills 129
Cheyenne Belt 94
Cheyenne Tableland 277
Chipata 332
Church Stretton 95
Cicxulub 194
Cinnabar 24
Clack, Jennifer 125
Clastic rocks 30
Cleveland Ironstone 176
Clypeus 181
Coal Measures 117, 146, 181
Coal, Australian 320
Coccoliths 187
Coelacanth 123
Coelophysis 164
Columbia River Plateau 195

Confuciusornis 190
Congo River/Lualaba 281
Conodont 106
Consequent River 276
Coombe Rock 251
Cope, Edward 183
Copper ore formation 330
Copperbelt, Zambia 330
Coral reef 264
Corallian Crag 223
Corallian Limestone 176, 178
Coriolis Force 57
Cornbrash 176, 177
Crag and Tail 256
Craven Fault 142
Creswell Crags 241
Cretaceous Interior Seaway 189, 197
Cretaceous Marine Transgression 188
Crimea 338
Crinoid 135
Cripple Creek 323
Crocodile 205
Cueva de los Cristales 21
Cyanobacteria 80, 87
Cyclothem 136
Cynodont 164
Dalradian formation 91
Darling Fault 317
Dartmoor 131
Darwin, Charles 11, 12, 14, 15, 264
Davis William Morris 272
De la Beche 180
De la Beche, Henry 122

Death Valley 296
Deccan Traps 194
Deflation Hollow 267
Deinotherium 214
Denisovan Man 233
Derbyshire Peak District 135, 144
Devensian Ice Age 235
d'Halloy, Jean 187
Diamond 23
Dickinsonia 93
Dicrodium 321
Didymograptus 108
Dimethyl sulphide 229
Dimetrodon 157
Dinosaur 164
Diplodocus 174, 183
Diprotodon 223, 328
Dob's Lin 103
Dolerite 27, 32
Dolomite 32
Donets Basin 336
Downs 217, 251
Dropstones 257
Drumlin 254
Duck-billed platypus 165, 188 328
Dungeness 262
Dunwich 242
Dyke 28
East African Rift Valley 78
East Pacific Rise 44
Echidna 165, 188
Echinoid 181, 187
Eden Project 171

Ediacaran fauna 92, 316
Edinburgh Castle 256
El Nino 240
Elbe Gorge 191
Elland Flags 142
Emerald 23
English Channel, formation of 233, 234
Entelodont 204
Eocene 202
Eocene Optimum 202
Eoraptor 164
Erosion 245
Erratics 242, 257
Essay on the Principles of Population 15
Estuarine Series, Yorks 177
Ethiopia Early Man 232
Etruria Marl 152
Études sur les glaciers 12
Eukaryotes 80, 84
Euramerica 152
Evaporites 166, 167, 213
Exhumation 277
Face of the Earth 162
Farallon Plate 295, 297, 298
Farnham Gap 275
Fayum 205
Feldspar 20
Ferrar Dolerite 321
Fingal's Cave 207
Finland 16-19
Flamborough Head 192
Flint 32
Florida Geology 265

Flysch 214
Foraminiferans 187, 198
Ford, John 269
Formicium giganteum 205
Fuller's Earth 177
Gabbro 23
Gabbro 24, 247
Gaia 54, 55
Galapagos 15
Galena 24
Ganges 220
Gastropod 98
Gatornis 198
Gault Clay 193
Gawler 314
Geike, Archibald 114
Geological Map of Southern Britain 182
Geological Map, Bath 181
Ghost Ranch 164
Giant's Causeway 207
Ginkgo 175
Glarus 214
Glen W 46
Glossopteris 38, 158, 163
Glyptodont 222, 237
Gneiss 32, 86
Gold (Welsh) 101
Gold 23
Gomphothere 214
Gondwana 104, 152, 154, 163, 317, 322
Goreme 200
Gough's Cave 241
Gould Stephen Jay 99

Graben 75
Grand Canyon 270
Grand Canyon of Yosemite 289
Grand Tetons 296
Granite (SW England) 167, 168, 178
Granite 20, 23, 24
Granite batholiths 94
Granite Controversy 170
Granite Emplacements, Australia 318
Granite, Cairngorms 113
Granites, Canberra 318
Graptolites 102, 103
Grass 211, 213
Great Artesian Basin 323
Great Devonian Controversy 123
Great Dividing Range 287, 325, 326
Great Glen Fault 113
Great Lakes 255
Great Oolite 177
Great Orme 135
Greenland Ice Cores 231
Greensand, Lower 193
Greensand, Upper 193
Grenville Series 88
Greywacke 31
Greywethers 251
Gryphaea 177
Gubbio 194
Guilin Limestone 260
Gulf of Bothnia 237
Gulf Stream 61, 221
Gwna Melange 93
Gymnosperm 134, 190

Gypsum 20
Hadley cell 57
Hadrian's Wall 140
Hadrocodium 175
Haematite 24
Half Dome 254
Hallucigenia 99
Halophytes 264
Hampshire Basin 206
Hardrosaur 190
Harz Mountains 153
Hawaii 68-70
Hebridean Tertiary Volcanics 207
Heezen, Bruce 41--3
Hercynian Orogeny 152
High Force 140
Himalayan Mountain Building in China 303
Himalayas 213, 220, 221
Hoffman, Paul 88
Holderness Coastal Erosion 262
Holmes, Arthur 49, 170
Holocene 237
Holzmaden 172
Homo erectus 232
Homo habilis 232
Homo heidelbergensis 236
Homo sapiens 232, 236, 238
Hoodoo 200
Hornblende 21
Horse Ice Age Survival 237
Horseshoe crab 107
Horst 75
Hoxnian Interglacial 235

Hsanda Gol 204
Humber Spit 242
Hutton James 9, 115
Huxley, Thomas 15, 16
Hwang He 308
Iapetus Ocean 101, 115
Ice Age 225
Iceland 64-6
Ichthyosaur 164, 174
Ichthyosaurus 180
Ichthyostega 124-6
Ignimbrite 28
Iguanodon 180
Iguanodon 180, 193
Indricothere 211
Indus 220
Ingleborough 248
Ingleborough Caves 258
Inselberg 248
Insubric Line 216
Inter-Tropical Convergence Zone 57
Ipswichian Interglacial 235
Irish Elk 237
Irtysh 337
Isle of Wight 193
Jarvik 125
Java Man 232
Jehol Lagerstätte 190
Joachimstal 24
Johanson, Don 232
Johnstone, Sir John 182
Jungfrau 216
Jura 217, 276

Jurassic Coast Dorset 172, 178
Jurassic Park 190
Kalgoorlie 23, 314
Kam 256
Kamchatka 335, 336
Kaolinite 171
Karelia 335
Katangan Supergroup 330
Kelvin, Lord 15
Kerguelen 188
Keuper Marl 166
Kilimanjaro 78
Kimmeridge Clay 176, 178
Kinder Scout 142
Knick Point 272
Knock and Lochan 252
Kobe 67
Kola Peninsula 335
Koonwarra Lake Deposit 324
Kopje 248
K-T Boundary 194-6
Kuznetsk Basin 336
Labirinthodont 321
Laetoli 232
Lagerstätten 99, 172, 205
Lake District Drainage 278
Lake Eyre 287
Lapworth, Charles 102, 103, 114
Laramide Orogeny 188, 295
Larvikite 24
Laterite 248
Laurentia 101, 109, 187
Leh 218

Lena 283, 337
Lepidodendron 147
Lewisian Gneiss 90
Liaoning 172
Lignin 146
Limestone Caves 258
Lindisfarne 139
Liopleurodon 173
Lithosphere 20
Livingstone Falls 281
Livingstone, David 281
Lobefish 124
Loess 251
London Basin 206
London Clay 193, 205
London Geology (West) 206
Longmynd 93
Lonsdale, William 123
Lord Howe Rise 324
Lovelock, James 54-56
Luangwa Valley 280, 331
Lucy 232
Lulworth Cove 178
Lungfish 124
Lycopod 147
Lyell, Charles 11, 12
Lyme Regis 180
Madagascar 188
Magnesian Limestone 162
Magnetite 24
Magnetometer 41
Malaspina Glacier 255
Malawi, Lake 79

Mallam Cove 258
Malthus, Thomas 15
Mammal Evolution 198
Mammal fossils 204
Mammoth 232
Mantell, Gideon 180
Marble 33
Mariana Trench 43
Marine Isotope Stages 235
Marine Transgressions 189
Marlborough Downs 192
Marsh, Othniel 183
Matterhorn 254
Maunder Minimum 241
McLeod Gamj 219
Mechanical erosion 245
Medieval Warm Period 241
Mediterranean Sea 213
Megalania 328
Megalosaurus 174
Meganaura 146
Megatherium 222
Megatherium 222, 237, 238
Mendip Hills 135
Mercian Mudstone 166
Mesa 268
Messel Shale 205
Meteor Strike at K-T boundary 194
Methane 21
Methane 54
Mica 20
Microgranite 209
Microraptor 190

Mid-Atlantic Ridge 39, 41
Midcontinental Rift 95
Migmatite 32
Milankovitch Cycles 226, 231
Milford Haven 262
Miller, Stanley 80
Millstone Grit 141, 185
Miocene 213
Misfit 273
Missoula, Lake 243
Mistaken Point 93
Moh's Scale 34
Moine formation 91
Moine Thrust 113
Monadnock 248
Monian Group 93
Monkey Puzzle 134
Monotreme 164
Mons Olympus 43
Monument Valley 268, 269
Moon 80
Moraine 256
Moran, Thomas 289
Morgan, Elaine 233
Morozova, Valentina 199
Morrison Formation 172, 183
Mount Isa 314
Mount Monadnock 248
Mount Pelee 71
Mount St Helens 67
Mull 207
Murchison, Roderick 101, 114, 123, 157
Murmansk 338

Murray Basin 325
Murray-Darling 287
Myllocunminga 100
Nappe 110
Neanderthal Man 233, 236
Nevadan Orogeny 294
New Forest 206
New Red Sandstone 165, 166
New York geology 111, 112
Niagara Falls 273
Nile 282
Norfolk Broads 242
North York Moors 177
Norwich Brickearth 242
Nothofagus 328
Ob 283, 337
Obsidian 24
Occam's Razor 12
Odonata 157
Old Faithful 289
Old Man of Hoy 128
Old Red Sandstone 110, 128
Oldoinjo Lengai 78
Oligocene 210
Olistostrome 31
Olivine 21
On the Nature of Limbs 14
Oolitic Limestone 176, 177
Ophiolite 48
Ophiolite 48, 293
Oregon Trail 278
Origin of the Species 14
Ornithiscians 175

Orogeny 110
Orthoclase 23
Orwell, George 149
Ouachita Mountains 152
Owen, Sir RICHARD 14
Oxford 177
Oxford Clay 176, 177
Oxidation 245
Oxygen Isotope Stages 235
Pacific Ring of Fire 67
Pakefield Stone Assemblage 236
Palaeoclimate 50
Paleocene 197
Paleocene-Eocene Thermal Maximum 202
Panama, Isthmus of 222
Pangaea 157, 187
Pangolin 204
Paradoxides 100
Parallel Roads of Glenroy 256
Peabody Museum 183
Peach and Horn 114
Pegmatite 24
Peking man 232
Peneplain 272
Pennant Sandstone 148
Pen-y-Ghent 248
Peridotite 28
Periglacial Landforms 250
Permian Basin, Texas 157
Permian Extinction 159, 160
Permo-Carboniferous Glaciation 154
Petigurus 115
Phanerozoic 84

Philadelphia Academy of Natural Science 183
Phillips, John 182
Phlegrean Fields 73
Piedmont Glacier 255
Piedmont, Desert 267
Pilbara 314, 316
Placental Mammal 165
Plagioclase 23
Pleistocene 225
Plesiosaur 163
Plesiosaur 163, 180
Plinean eruption 71
Pliocene 221
Pluton 24
Plymouth 262
Podocarp 175
Porphyry 20
Port Askaig tillite 91
Portland Bill 179
Portland Stone 176, 179
Potosi 23
Poundstone 180
Powell, John Wesley 271 277
Precession of the Equinoxes 226
Principles of Geology 11
Principles of Physical Geology 49
Prokaryotes 80
Propaleotherium 205
Proterozoic Australia 314
Protoaxites 127
Pterodactyl 180
Pterosaur 164, 190

Pyroclast 28
Pyroxene 21
Quartz-dolerite dykes 209
Quartzite 33
Quetzacoatlus 190
Radiolarians 187
Rasmussen, Steen 81
Ray-finned fish 123
Reading Beds 193, 205
Red Crags 223
Red Cuillins 209
Red Lady of Paviland 180, 236
Redbeds, Sydney 321
Reptiles 17
Resequent River 276
Reynolds, Doris 170
Rhaetic 162
Rhoetosaurs 322
Rhynie Chert 127
Rhyolite 27, 28
Ria 262
River at the Centre of the World 305
River Capture 273
Riversleigh 212, 327
Road to Jaramillo 46
Road to Wigan Pier 149
Roches moutonnées 12
Rocky Mountains 188, 277, 293
Rodinia 88
Romney Marshes 242
Rotunda Scarborough 181
Salima 333
Salisbury Plain 192

San Andreas Fault 77
San Miguel Azores 186
Santorini 74
Sapphire 23
Sarsen Stones 198
Saurischians 175
Sauropod 164, 183
Schist 32
Scourie Dykes 90, 91
Sea floor spreading 44-46
Sedgwick, Adam 15, 97, 102, 123
Selenite 20
Serpentinite 48
Sever Orogeny 294
Shale bing 151
Shark 135, 198
Shocked Quartz 194
Shubin, Neil 126
Siberia, Western 335
Siberian Rivers 283
Siberian Traps 160
Siccar Point 115
Sierra Nevada 294, 298, 299
Sigillaria 157
Sill 28
Sink Hole 257
Skiddaw Group 117
Skye 207
Slate (Welsh) 101
Slave Province 94
Slimonia 126
Smilodon 223
Smilodon 223, 236

Smith 180
Smith William 10
Snake River Basin 195
Snowball Earth 88, 809
Snowball Earth, Australia 316
Snowdon 107
Solfatara 73
Solifluxion 251
Solnhofen 172
Somersetshire Coal Canal Company 181
South American Fauna 188
South Wales Coalfield 148
Spartina townsendii 264
Sphalerite 24
Sprigg, Reg 92
Spriggiana 92, 93
St Paul's Cathedral 179
Staffa 207
Stanage Edge 142
Stanley, Henry 281
Steeno, Nicholas 9
Stegosaurus 174, 183
Steinman Trinity 48
Stirling Castle 140
Stone Polygons 251
Stonehenge 198
Stromatolite 86
Subbotina, Nina 199
Subduction 44
Subsequent River 276
Suess, Eduard 73, 162
Superimposed Drainage 277
Superior Province 94

Syenite 24
Synapsid 158
Syncline 108
Taconic Orogeny 111, 290
Takla Makan 306
Tamyr Peninsula 336
Tangshan Earthquake 304
Teal J H H 90
Terror Bird 198
Tethys Ocean 197, 213, 306
Thames Valley Ice Age Fauna 235
Tharp, Mary 41-3
The Day After Tomorrow 231
Theory of the Earth 9
Therapsid 159
Three Gorges Dam 311, 312
Thylacoleo 328
Tiktaalik 126
Titanites 179
Titanoboa 198
Tor 132
Toridonian Sandstone 91
Toxodon 237
Trees, Broad-leaved 203
Triceratops 175
Triceratops 175, 190
Trilobite 97, 98
Trilobites Beetle Creek, Australia 317
Trypanosomiasis 331
Tsavo National Park 248
Turbidite 31
Tyrannosaurus Rex 190
Unifomitarianism 11

Urey, Harold 51, 80
U-Shaped Valley 253
Vale of Pickering 242
Variscan Orogeny 152
Varve 257
Verrucano 214
Vesuvius 28, 70
Victoria Falls 281
Vine and Matthews paper 46
Vladivostok 338
Volga River 337
Walcott, Charles 99
Wallace Line 214
Wallace, Alfred 214
Weald 192
Weald Anticline 275
Weald Clay 192
Wealden Facies 193
Weathering 245
Wedgwood Josiah 152
Wegener, Alfred 40 41
Westerwald 153
Whin Sill 140
Whitby 176
Whitsunday Islands 323
Wigan 150
Wilberforce, Samuel 16
Wilson Tuzo 46
Winchester, Simon 305
Windermere Group 118
Wollaston Medal 182
Wolstonian 235
Wonderful Life 99

Xi River 310
Yangtze River 305
Yellowstone River 289
Yenisei 283
Yenisei 283, 337
Yilgarn 314
Yoredale Cyclothem 136
York Minster 166
Yorkshire Dales 135, 258
Yorkshire Rivers 285 286
Yorkshire Wolds 192
Yosemite 253
Young, James 149
Younger Dryas 239
Zambezi 298, 330
Zircon 50

www.ingramcontent.com/pod-product-compliance
Lightning Source LLC
Chambersburg PA
CBHW060821170526
45158CB00001B/42